A World of Mathematics

Book 4

G. Marshall

Head of the Mathematics Department
Ellers High School, Doncaster

Nelson

© G. Marshall 1971   First published 1971

Fourth impression 1976

ISBN 0 17 431219 9

Thomas Nelson and Sons Ltd
Lincoln Way Windmill Road
Sunbury-on-Thames Middlesex TW16 7HP
P.O. Box 73146 Nairobi Kenya

Thomas Nelson (Australia) Ltd
19-39 Jeffcott Street West Melbourne Victoria 3003

Thomas Nelson and Sons (Canada) Ltd
81 Curlew Drive Don Mills Ontario

Thomas Nelson (Nigeria) Ltd
8 Ilupeju Bypass PMB 1303 Ikeja Lagos

Designed by Melvyn Gill and Chris Strong
Illustrations by Peter Brooks

Filmset by Keyspools Ltd., Golborne, Lancs

Printed in Hong Kong

Contents

Introduction

A **World of Mathematics** is designed for pupils in secondary schools advancing at a moderate pace. Whilst providing a systematic treatment of the basic skills, as the title implies, the course embraces a wide range of mathematical topics. The work has been arranged in six books for two reasons: first, to make it easier to match the rate of progress to the ability of the children concerned; second, to provide a sense of achievement and the stimulus of a new book at more frequent intervals. The chapters are arranged in groups, usually including number, measurement and spatial relationships, and this constant variation in topics helps to sustain interest. Each group is followed by a revision exercise, since the inability to retain facts and ideas is one of the greatest problems encountered at this level. Indeed the need for constant revision cannot be too strongly emphasized: the existing 'edifice' of a topic should always be re-examined before an attempt is made to build on to it.

The remedial element in the first two books is continued, and the exercises on number have a diagnostic value which will reveal the need for supplementary examples.

There is a constant need to find ways of motivating pupils who find the subject difficult, and this was especially borne in mind when the series was planned. The relative simplicity of the number work provides for a reasonable degree of success which should engender confidence, there is an abundance of diagrams and drawings, and great care has been taken to produce an attractive page design. Whenever possible the topics are related to the everyday world within the child's experience, and the children are encouraged to collect illustrations and any other items related to their work. Several of the topics can be reinforced by the provision of assignment cards involving the school's own particular environment or drawn from information currently available.

Many ideas and suggestion are given in the books and the teacher's notes, but the individual teacher will fashion the contents to his own design.

Acknowledgments for photographs appearing in this book are due to: B.O.A.C. (p. 27), Camera Press (p. 27), J. Allan Cash (p. 50), the International Wool Secretariat (p. 41), Sport and General Press Agency Ltd. (p. 16), the United States Information Service (pp. 41, 42 and 57).

Sets

Look carefully at these two sets.

A = {1, 3, 5, 7, 9, 11}
B = {3, 6, 9, 12}

A is the set of odd numbers between 0 and 12.
B is the set of multiples of 3 between 2 and 13.
When we compare the two sets we find that the numbers 3 and 9 are members of both of them.
Let us see how this could be shown by diagrams.
Here is Set A on its own.   Here is Set B on its own.

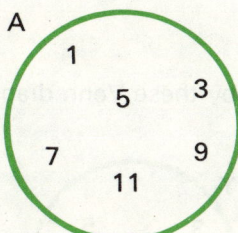

Here are the two sets
overlapping so that
each contains the same
members as before.

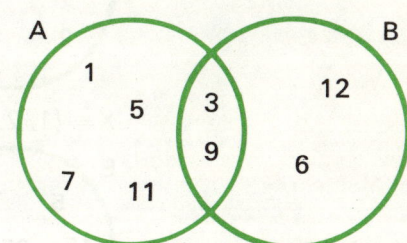

In the language of sets we say that the two sets **intersect** and you will see that the **intersection** of the two sets is made up of the members that belong to both.
The kind of picture we have used is called a **Venn** diagram after John Venn, the Scottish mathematician who thought of the idea.

This is the sign we use for
the intersection of sets. $\cap$

So when we write A $\cap$ B = {3, 9} we mean that the intersection of the two sets A and B is 3, 9.
Here is another example.
If X = {2, 4, 6, 8, 10} and Y = {1, 2, 3, 4, 5}, then the intersection of X and Y is written like this:

X $\cap$ Y = {2, 4}

Look at these two sets.

C = {△ □ ⬠} D = {0 ⬭ 0}

They have no members in common, so the intersection is said to be an empty set.

Here are the two sets shown as Venn diagrams. Since they do not have any members in common we do not show them overlapping.

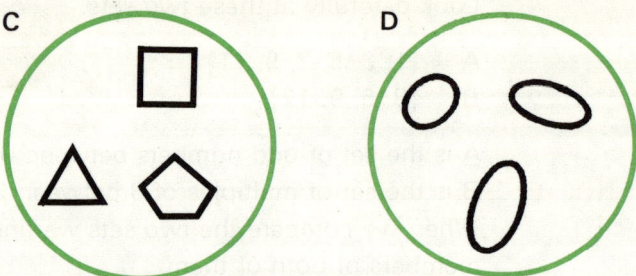

In such cases we write C ∩ D = ∅.

Sets which do not have any members in common are called **disjoint** sets.

Exercises

A List the sets shown by these Venn diagrams.

Example:

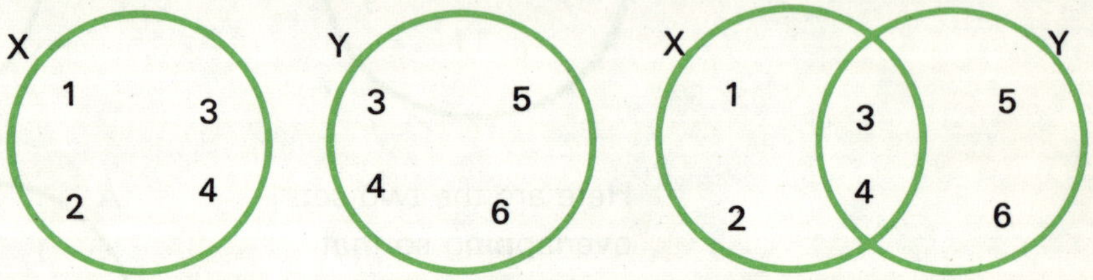

X = {1, 2, 3, 4}   Y = {3, 4, 5, 6}   X ∩ Y = {3, 4}

1

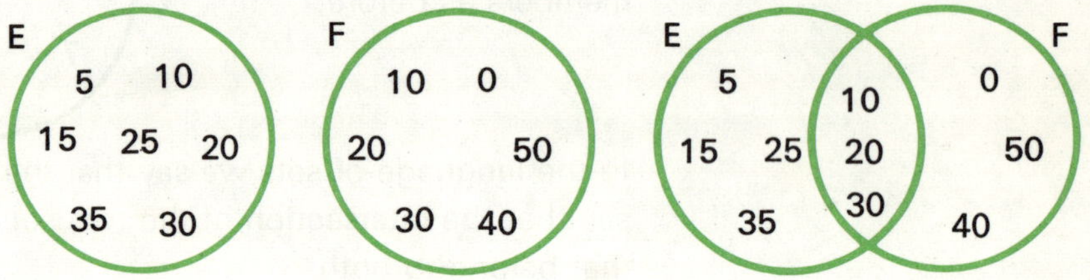

E: 5, 10, 15, 25, 20, 35, 30
F: 10, 0, 20, 50, 30, 40
E: 5, 15, 25, 35; 10, 20, 30; F: 0, 50, 40

2

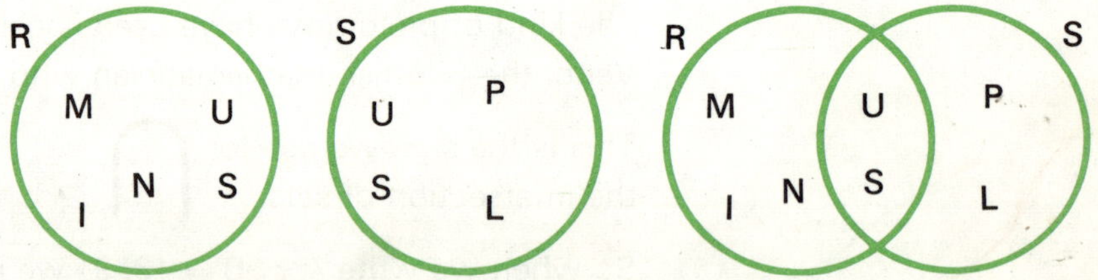

R: M, U, I, N, S
S: U, P, S, L
R: M, I, N; U, S; S: P, L

3

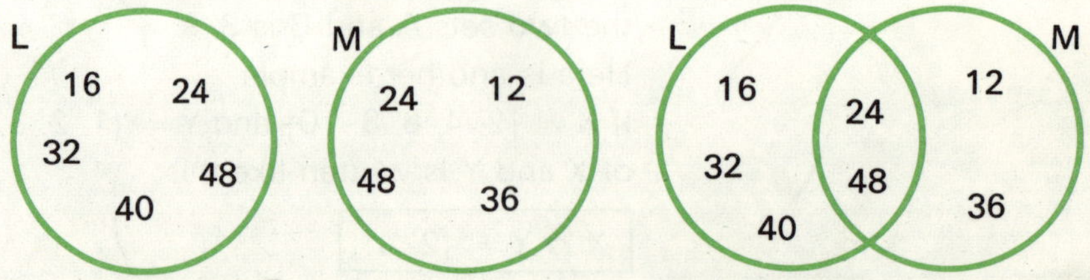

L: 16, 24, 32, 48, 40
M: 24, 12, 48, 36
L: 16, 32, 40; 24, 48; M: 12, 36

4

J: 7, 14, 21, 49, 42, 28, 35
K: 12, 36, 18, 42, 6, 24, 30, 48
J: 7, 14, 49, 21, 42, 28, 35; 36, 6, 30, 48, 24; K: 12, 18

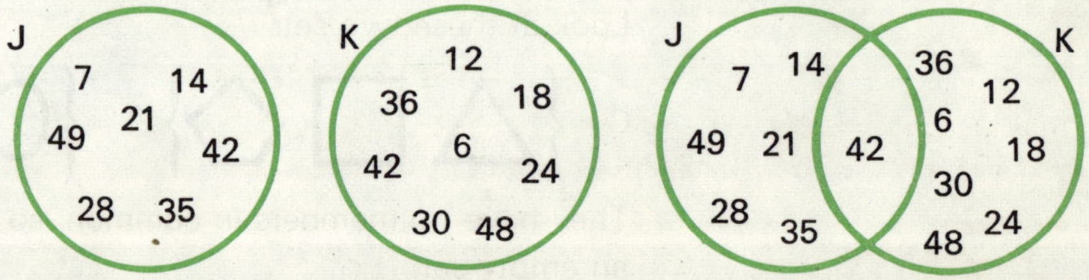

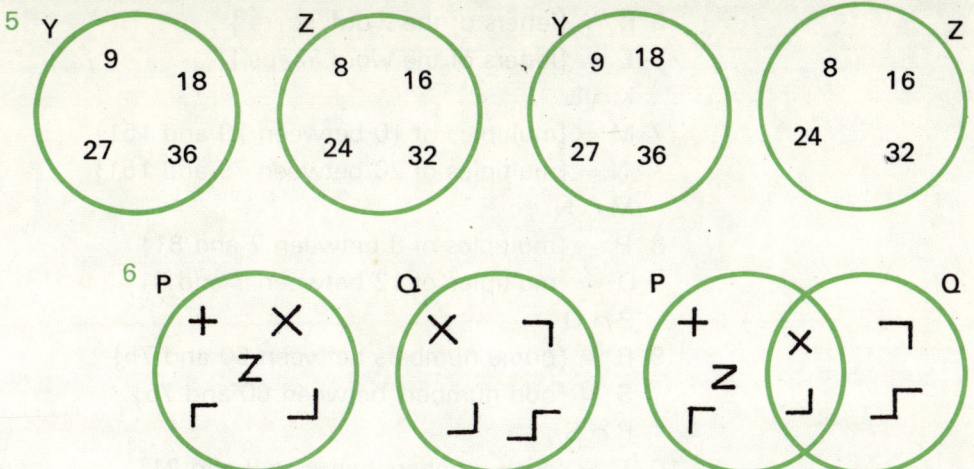

B (a) Draw Venn diagrams to show the following sets and intersections.

(b) List the intersections.

1   A = {1, 10, 100}, B = {10, 100, 1,000}, A ∩ B.
2   C = {2, 4, 6, 8, 10, 12}, D = {3, 6, 9, 12, 15}, C ∩ D.
3   E = {b, d, f, h, k, l, t}, F = {g, j, p}, E ∩ F.
4   G = {11, 22, 33, 44, 55}, H = {22, 44, 66, 88, 110}, G ∩ H.
5   I = {triangle, quadrilateral, pentagon}.
    J = {quadrilateral, rectangle, square, parallelogram}, I ∩ J.
6   K = {$\frac{1}{4}$, $\frac{1}{2}$, $\frac{1}{5}$, $\frac{1}{3}$}, L = {$\frac{1}{7}$, $\frac{1}{5}$, $\frac{1}{4}$, $\frac{1}{6}$}, K ∩ L.
7   M = {0·1, 0·01, 0·001}, N = {0·01, 0·05, 0·03, 0·07}, M ∩ N.
8   P = {6, 12, 18, 24, 30, 36}, Q = {9, 18, 27, 36, 45}, P ∩ Q.
9   R = {octave, octopus, October, octagon},
    S = {heptagon, octagon, decagon, pentagon}, R ∩ S.
10 T = {$\frac{1}{2}$p, 1p, 2p, 5p, 10p, 50p},
    Y = {5p, 10p, 15p, 20p}, T ∩ Y.

C Write out the members of the following sets and intersections.

1   A = {prime numbers between 10 and 30}
    B = {odd numbers between 14 and 30}
    A ∩ B
2   C = {square numbers between 3 and 37}
    D = {multiples of 4 between 3 and 37}
    C ∩ D
3   E = {odd numbers between 94 and 110}
    F = {multiples of 12 between 94 and 110}
    E ∩ F
4   G = {vowels of the English alphabet}
    H = {letters of the word 'aisle'}
    G ∩ H
5   I = {square numbers between 20 and 50}
    J = {prime numbers between 20 and 50}
    I ∩ J

6 K = {letters of the word 'spare'}
   L = {letters of the word 'reaps'}
   K ∩ L

7 M = {multiples of 10 between 79 and 151}
   N = {multiples of 20 between 79 and 151}
   M ∩ N

8 P = {multiples of 8 between 7 and 81}
   Q = {multiples of 12 between 7 and 81}
   P ∩ Q

9 R = {prime numbers between 50 and 75}
   S = {odd numbers between 60 and 75}
   R ∩ S

10 T = {even numbers between 9 and 21}
   V = {multiples of two between 9 and 21}
   T ∩ V.

D Look at these sets and then say whether the following statements are true or false.

A = {3, 6, 9, 12} D = {2, 4, 6, 8, 10, 12}
B = {4, 8, 12, 16, 20} E = {1, 4, 9, 16, 25}
C = {7} F = {9, 3, 6, 12}

1 A ∩ E = 9
2 A = F
3 A and C are disjoint sets
4 B ∩ D = {4, 8, 10}
5 B ∩ C = ∅
6 E ∩ F = {12}

7 A ∩ D = ∅
8 D ∩ F = {6, 12}
9 B ∩ E = {4, 16}
10 A ∩ B ∩ D = {12}
11 A ∩ E = E ∩ A
12 A ∩ C = B ∩ C

Multiplication

Alan has three lots of marbles and Ian has two. There are eight marbles in each lot. How many marbles have they together?

Alan   *Ian*

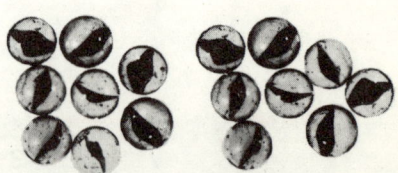

We could say

either   **or**

Alan has 3 × 8 = 24 marbles
Ian has 2 × 8 = 16 marbles
Together they have
24 + 16 = 40 marbles.

Together they have
3 + 2 = 5 lots
Together they have
5 × 8 = 40 marbles.

You will see that in the first method there are three operations whereas in the second method there are only two. The second method is therefore quicker and easier, and so we shall practise using this method. This is how we set out the working:

$$(3 \times 8) + (2 \times 8) = (3 + 2) \times 8$$
$$= 5 \times 8$$
$$= 40$$

Exercises

A   Find the total number of dots in each of the following arrays. Set out your working as shown in the example above.

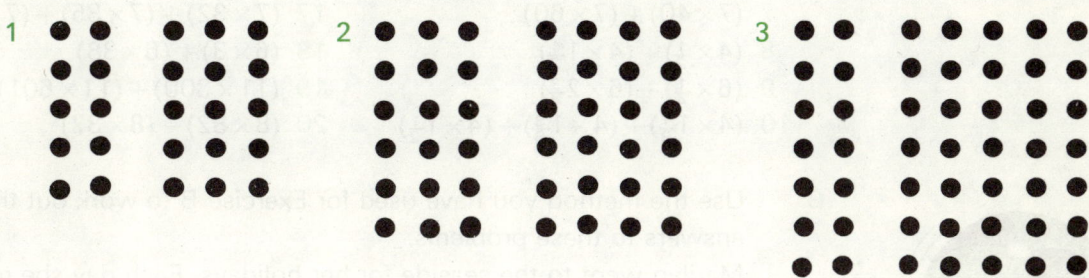

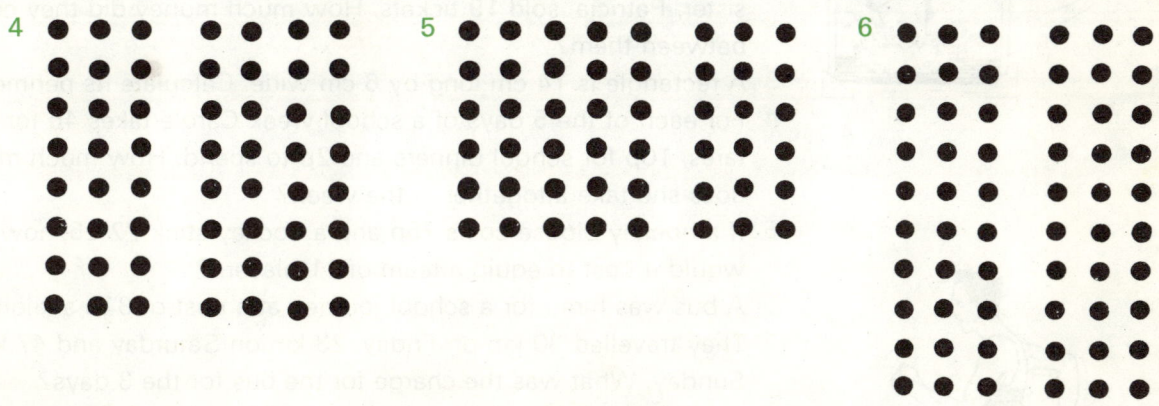

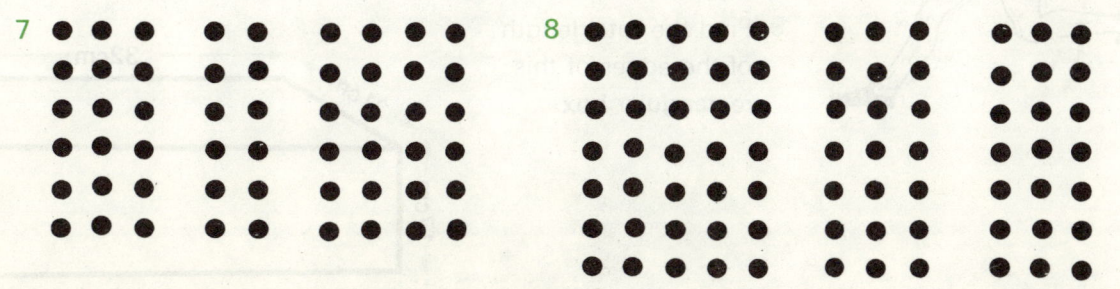

B You will perhaps remember that when we are multiplying, the order of the numbers may be changed, for example: $4 \times 5 = 20$ and $5 \times 4 = 20$. In this exercise notice that the number is placed first and the bracket last.

Example: $(4 \times 15) + (4 \times 25) = 4 \times (15 + 25)$
$$= 4 \times 40$$
$$= 160$$

1 $(3 \times 2) + (3 \times 8)$
2 $(5 \times 7) + (5 \times 3)$
3 $(4 \times 3) + (4 \times 8)$
4 $(2 \times 7) + (2 \times 13)$
5 $(9 \times 4) + (9 \times 8)$
6 $(4 \times 2) + (4 \times 3) + (4 \times 4)$
7 $(7 \times 40) + (7 \times 60)$
8 $(4 \times \frac{1}{2}) + (4 \times 1\frac{1}{2})$
9 $(6 \times \frac{1}{4}) + (6 \times 2\frac{3}{4})$
10 $(4 \times 1\frac{1}{8}) + (4 \times 1\frac{3}{8}) + (4 \times 1\frac{1}{2})$

11 $(3 \times 998) + (3 \times 2)$
12 $(9 \times 5) + (9 \times 7) + (9 \times 8)$
13 $(\frac{1}{2}$ of $13) + (\frac{1}{2}$ of $17)$
14 $(12 \times 24) + (12 \times 26)$
15 $(14 \times 47) + (14 \times 3)$
16 $(8 \times \frac{2}{3}) + (8 \times 4\frac{1}{3})$
17 $(7 \times 32) + (7 \times 35) + (7 \times 33)$
18 $(6 \times 3) + (6 \times 38)$
19 $(11 \times 300) + (11 \times 601) + (11 \times 99)$
20 $(8 \times 82) - (8 \times 32)$

C Use the method you have used for Exercise B to work out the answers to these problems.

1 Marilyn went to the seaside for her holidays. Each day she received 12p from her father and 8p from her grandfather. How much spending money did she get if she was away for 8 days?

2 Tickets for a concert cost 9p each. Susan sold 11 tickets and her twin sister, Patricia, sold 19 tickets. How much money did they collect between them?

3 A rectangle is 14 cm long by 6 cm wide. Calculate its perimeter.

4 For each of the 5 days of a school week Carole takes 4p for her bus fares, 10p for school dinners and 2p to spend. How much money does she take altogether in the week?

5 If a hockey blouse costs 75p and a hockey stick £2·25, how much would it cost to equip a team of 11 players?

6 A bus was hired for a school journey at a cost of $8\frac{1}{2}$p a kilometre. They travelled 30 km on Friday, 23 km on Saturday and 47 km on Sunday. What was the charge for the bus for the 3 days?

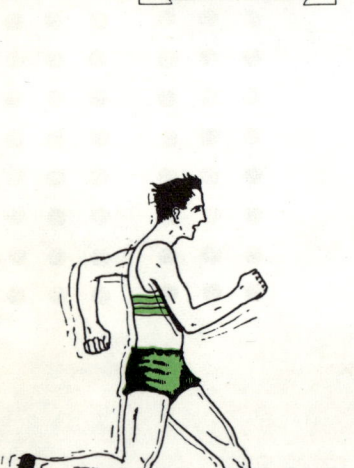

7 On 5 days of the week an athlete did 40 minutes training before breakfast, 25 minutes in the middle of the day, and 55 minutes in the evening. How much time did he spend on training throughout the week?

8 Find the total length of the edges of this rectangular box.

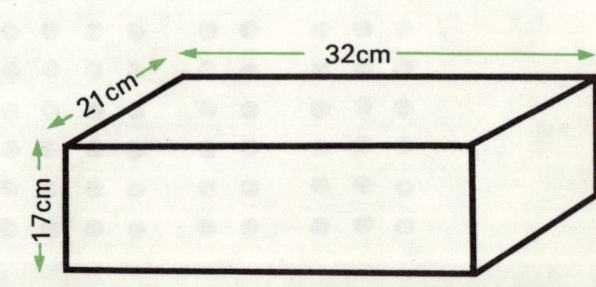

D Work out the following, using an easy way if you can see one.

1	2 × 12 × 15	8	36 × 88	14	20 × 60 × 70
2	9 × 18	9	23 × 64	15	9 × 999
3	50 × 7 × 2	10	35 × 123	16	16 × 51
4	25 × 44	11	25 × 6 × 4	17	6 × 80 × 5
5	7 × 99	12	10 × 12 × 100	18	101 × 101
6	14 × 32	13	2 × 6 × 7	19	5 × 21 × 8
7	22 × 50			20	46 × 238

3 Measuring Distance

We have already learned that for measuring length the standard unit is the metre.

For longer distances we use the kilometre.

Here are some important facts about the kilometre (km).

1 Kilo stands for the number 1,000.

2 1 kilometre is equal to 1,000 metres.

3 A kilometre is about four times round a soccer pitch   two and a half times round an athletics track.

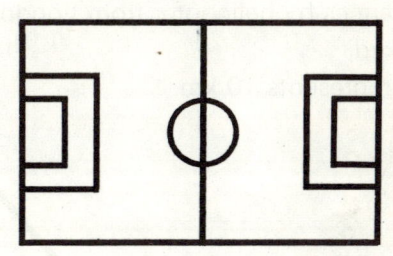 or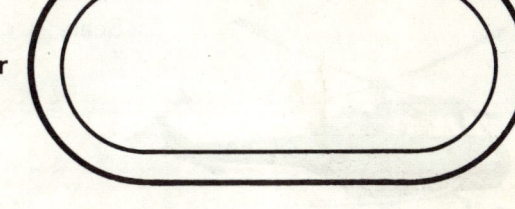

4 A man walks at a speed of about 6 kilometres an hour.

5 A normal speed limit for traffic in towns is 50 kilometres an hour.

To change kilometres to metres we multiply by 1,000. We do this by moving each figure 3 places to the left. Notice that we must put noughts in any whole number places that are left empty. For whole numbers we put three noughts to the right of the number.

			1·5	kilometres

equal

1	**5**	**0**	**0·**	metres

To change metres to kilometres we divide by 1,000. We do this by moving the figures 3 places to the right.

4	**3**	**0**	**0·**	metres

equal

		4·3	kilometres

A Find the distances represented by these lines. Express the parts of a kilometre in decimals. **Scale:** 1 cm rep. 1 km.

Example:

——————————————————————— 5·3 km

1 ———————————————————————

2 ————————————————————————————

3 —————————

4 ——————————————

5 ——————————————————————————————————

6 ————————————————————————————————

7 ——————————————————————————————

8 ———————

9 ——————————————————————

10 ———————————————————————————————————————

B Find the distances by helicopter from London Heliport to the airfields marked.
Scale: 1 cm represents 10 km.

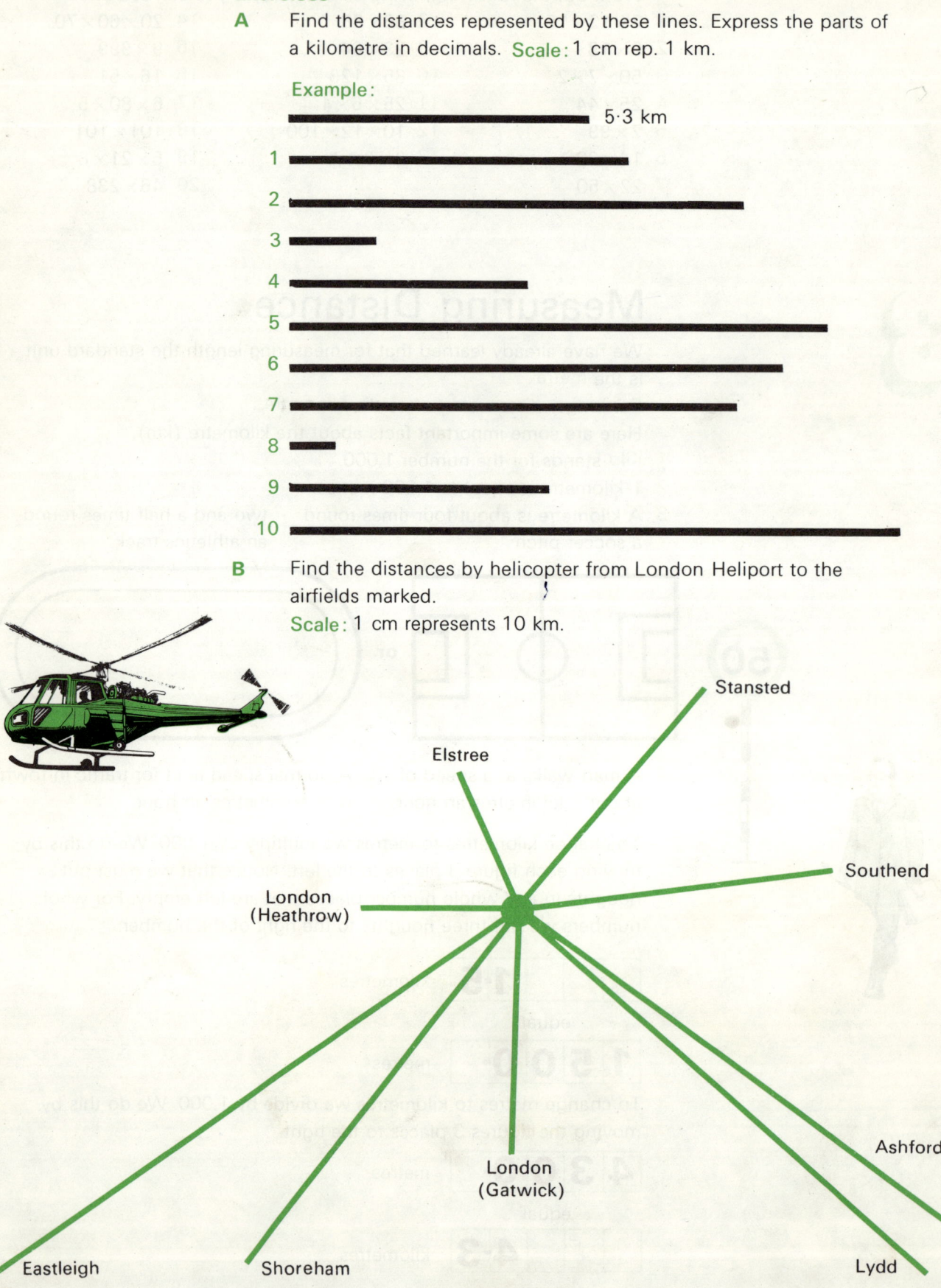

12

C For this exercise you will need a large scale map of your own district.

1 Draw a circle radius 5 cm.

2 In the centre of your circle put a cross to represent your school.

3 Mark in the four cardinal points of the compass.

4 Underneath your circle write: Scale: 5 cm represents 1 km.

5 From your map find a number of places which are about one kilometre from your school and mark them in as shown.

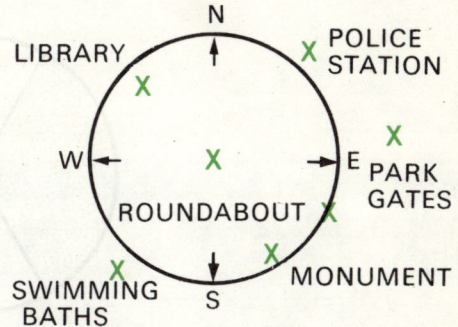

D Change to metres:

1	2 km	6	2·5 km	11	1·01 km	16	8·12 km
2	7 km	7	8·5 km	12	1·21 km	17	8·012 km
3	9 km	8	12·5 km	13	4·09 km	18	4·876 km
4	11 km	9	7·8 km	14	4·90 km	19	30·038 km
5	15 km	10	8·7 km	15	2·003 km	20	123·456 km

E Change to kilometres:

1	3,000 m	6	3,500 m	11	4,030 m	16	1,001 m
2	6,000 m	7	8,500 m	12	6,020 m	17	3,008 m
3	11,000 m	8	12,500 m	13	12,080 m	18	8,750 m
4	17,000 m	9	16,100 m	14	30,030 m	19	8,075 m
5	20,000 m	10	20,100 m	15	54,060 m	20	64,429 m

F In the following examples you are given the average speed and the time for a number of car journeys. Find the distance travelled in each case.

	Kilometres an hour	Time		Kilometres an hour	Time		Kilometres an hour	Time
1	40	3 h	6	49	7 h	11	20	$2\frac{1}{2}$ h
2	45	4 h	7	40	10 h	12	48	$3\frac{1}{2}$ h
3	52	5 h	8	39	8 h	13	60	$4\frac{1}{4}$ h
4	64	6 h	9	65	11 h	14	95	13 h
5	37	4 h	10	43	12 h	15	36	4 h 20 min

G How many of your own paces would it take to walk a kilometre? Do you know a quick way to find out?

Common Fractions

Equivalent Fractions

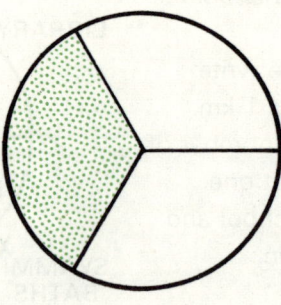 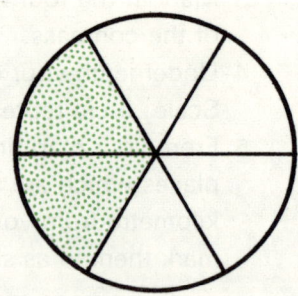 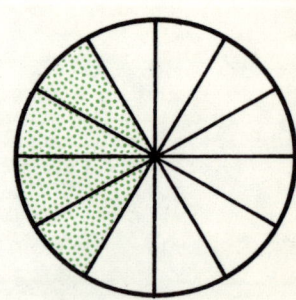

one third   two sixths   four twelfths

The above diagrams remind us:

(a) that **equivalent fractions** occupy the same part of a whole one,

(b) that if we multiply the numerator and denominator of a fraction by the same number the resulting fraction is equivalent to the original one.

For example: $\quad \frac{1}{3} = \left(\frac{1}{3} \times \frac{2}{2}\right) = \frac{2}{6} \qquad \frac{1}{3} = \left(\frac{1}{3} \times \frac{4}{4}\right) = \frac{4}{12}$

We often have to use this method for obtaining equivalent fractions when adding and subtracting.

Addition of Mixed Numbers

When we add mixed numbers the working is simplified by adding the whole numbers separately. Study the examples carefully.

$$2\tfrac{1}{5} + 3\tfrac{1}{2} = (2+3) + \left(\tfrac{1}{5} + \tfrac{1}{2}\right)$$
$$= 5 + \left(\tfrac{2}{10} + \tfrac{5}{10}\right)$$
$$= 5\tfrac{7}{10}$$

$$1\tfrac{5}{8} + 2\tfrac{3}{4} = (1+2) + \left(\tfrac{5}{8} + \tfrac{3}{4}\right)$$
$$= 3 + \left(\tfrac{5}{8} + \tfrac{6}{8}\right)$$
$$= 3 + \tfrac{11}{8}$$
$$= 3 + 1\tfrac{3}{8}$$
$$= 4\tfrac{3}{8}$$

Halves	1								2							
Quarters	1				2				3				4			
Eighths	1		2		3		4		5		6		7		8	
Sixteenths	1	2	3	4	5	6	7	8	9	10	11	12	13	14	15	16

Use the above diagram to do the following examples. It will help if you know that since $\frac{1}{4}+\frac{1}{2}=\frac{3}{4}$ and $\frac{1}{2}+\frac{1}{4}=\frac{3}{4}$

then $\frac{1}{4}+\frac{1}{2}=\frac{1}{2}+\frac{1}{4}$

1 $\frac{1}{2}+\frac{1}{8}$
2 $\frac{1}{2}+\frac{1}{16}$
3 $\frac{1}{4}+\frac{1}{8}$
4 $\frac{1}{4}+\frac{1}{16}$
5 $\frac{1}{2}+\frac{3}{8}$
6 $\frac{1}{2}+\frac{3}{16}$
7 $\frac{1}{2}+\frac{5}{16}$
8 $\frac{1}{2}+\frac{7}{16}$

9 $\frac{1}{4}+\frac{3}{16}$
10 $\frac{1}{4}+\frac{5}{16}$
11 $\frac{1}{4}+\frac{7}{16}$
12 $\frac{1}{4}+\frac{9}{16}$
13 $\frac{1}{4}+\frac{11}{16}$
14 $\frac{1}{8}+\frac{1}{2}$
15 $\frac{1}{8}+\frac{3}{4}$
16 $\frac{1}{8}+\frac{1}{16}$

17 $\frac{1}{8}+\frac{7}{16}$
18 $\frac{3}{8}+\frac{1}{4}$
19 $\frac{5}{8}+\frac{3}{16}$
20 $1\frac{5}{16}+1\frac{1}{2}$
21 $1\frac{3}{4}+\frac{3}{16}$
22 $2\frac{3}{16}+1\frac{3}{8}$
23 $3\frac{7}{16}+2\frac{1}{4}$
24 $2\frac{5}{8}+1\frac{1}{4}$

25 $\frac{1}{2}+\frac{1}{4}+\frac{1}{8}$
26 $\frac{1}{2}+\frac{1}{4}+\frac{1}{16}$
27 $\frac{3}{16}+\frac{1}{4}+\frac{3}{8}$
28 $\frac{1}{8}+\frac{3}{16}+\frac{1}{4}$
29 $1\frac{7}{16}+1\frac{1}{8}+1\frac{1}{4}$
30 $\frac{3}{8}+1\frac{1}{2}+2\frac{1}{16}$
31 $1\frac{3}{4}+2\frac{1}{8}+\frac{1}{16}$
32 $2\frac{5}{8}+\frac{3}{16}+1\frac{1}{4}$

B 1 Write down the set of natural numbers between 0 and 6.
  2 Make as many different proper fractions as possible using pairs of numbers from your set.
  3 Represent each fraction on a graph as shown.
  4 Use your graph to put your set of fractions in order of size, smallest first.

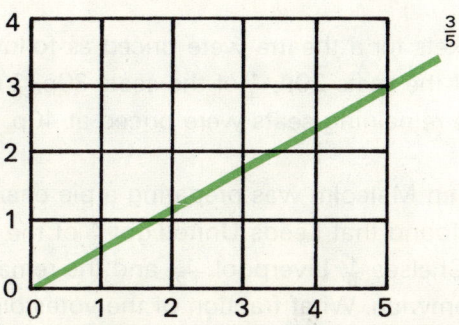

C   Example:

$$2\tfrac{1}{3}+4\tfrac{2}{5}$$

$$\tfrac{1}{3}=\tfrac{1}{3}\times\tfrac{5}{5}=\tfrac{5}{15}; \quad \tfrac{2}{5}=\tfrac{2}{5}\times\tfrac{3}{3}=\tfrac{6}{15}$$

$$2\tfrac{1}{3}+4\tfrac{2}{5}=(2+4)+\left(\tfrac{1}{3}+\tfrac{2}{5}\right)$$
$$=6+\tfrac{5}{15}+\tfrac{6}{15})$$
$$=6\tfrac{11}{15}$$

The number in the brackets after some of the examples tells you the common denominator you have to use.

1   $\tfrac{2}{3}+\tfrac{1}{6}$ 7 $\tfrac{1}{4}+\tfrac{3}{8}$ 13 $\tfrac{2}{5}+\tfrac{13}{15}$ 19 $\tfrac{1}{2}+\tfrac{2}{3}+\tfrac{1}{4}$

2   $\tfrac{2}{5}+\tfrac{3}{10}$ 8 $\tfrac{1}{2}+\tfrac{2}{7}$ 14 $5\tfrac{5}{8}+4\tfrac{7}{12}$ (24)   20 $1\tfrac{1}{5}+1\tfrac{1}{10}+1\tfrac{1}{15}$ (30)

3   $\tfrac{1}{4}+\tfrac{1}{3}$ 9 $\tfrac{3}{5}+\tfrac{7}{10}$ 15 $3\tfrac{7}{10}+5\tfrac{3}{4}$ (20)   21 $1\tfrac{1}{4}+2\tfrac{2}{5}+2\tfrac{3}{10}$ (20)

4   $\tfrac{1}{3}+\tfrac{3}{4}$ 10 $\tfrac{1}{3}+\tfrac{1}{5}$ 16 $4\tfrac{4}{9}+3\tfrac{2}{3}$ 22 $1\tfrac{1}{2}+2\tfrac{5}{6}+3\tfrac{7}{12}$

5   $\tfrac{2}{5}+\tfrac{1}{4}$ 11 $\tfrac{7}{10}+\tfrac{1}{3}$ 17 $4\tfrac{1}{3}+\tfrac{11}{12}$ 23 $3\tfrac{3}{5}+2\tfrac{7}{20}+4\tfrac{7}{10}$

6   $\tfrac{1}{6}+\tfrac{2}{3}$ 12 $\tfrac{3}{4}+\tfrac{3}{5}$ 18 $4\tfrac{9}{10}+2\tfrac{8}{15}$ (30)   24 $5\tfrac{4}{7}+\tfrac{5}{21}+4\tfrac{1}{3}$ (21)

D   1   Mr Chapman wanted to erect a fence around his garden. His daughter measured the garden and found it was $15\tfrac{1}{2}$ m long by $10\tfrac{3}{4}$ m wide. What length of fencing did Mr Chapman have to order?

2   Marlene's time on Tuesday morning was divided up as follows.

How long was the morning session?

Assembly $\tfrac{1}{2}$ h	Needlework $1\tfrac{1}{4}$ h	Break $\tfrac{1}{4}$ h	Mathematics $\tfrac{3}{4}$ h	English $\tfrac{3}{4}$ h

3   Tickets for a theatre were priced as follows: $\tfrac{1}{3}$ of the seats 20p, $\tfrac{1}{4}$ of the seats 30p, $\tfrac{1}{6}$ of the seats 50p. The remaining seats were priced at 40p. What was this fraction?

4   When Malcolm was preparing a pie chart on favourite football teams he found that Leeds United got $\tfrac{1}{3}$ of the votes, Manchester United $\tfrac{3}{8}$, Chelsea $\tfrac{1}{6}$, Liverpool $\tfrac{1}{12}$, and the remaining votes went to West Bromwich. What fraction of the votes did West Bromwich get?

John Hollins (Chelsea) in action at Stamford Bridge.

5 At a school sports the times for the 4×100 metres relay race were:

Blues	$12\frac{1}{5}$,	$12\frac{1}{5}$,	$12\frac{3}{10}$ and	$11\frac{4}{5}$ seconds
Greens	$11\frac{9}{10}$,	$12\frac{3}{5}$,	$12\frac{7}{10}$ and	$12\frac{1}{5}$ seconds
Reds	$12\frac{1}{2}$,	$12\frac{3}{10}$,	$12\frac{2}{5}$ and	$11\frac{9}{10}$ seconds
Yellows	$11\frac{7}{10}$,	$12\frac{4}{5}$,	$11\frac{9}{10}$ and	$12\frac{1}{2}$ seconds

Find the total time for each team and then place them in order.

Making Sure—One

A **1** Write down the set of numbers between 0 and 46 which can be divided exactly by 3 and by 5.

2 Write down the set of compass points which are exactly opposite to the members of the following set: {N, NE, E, SE}

3 Write down the sets illustrated by these Venn diagrams.

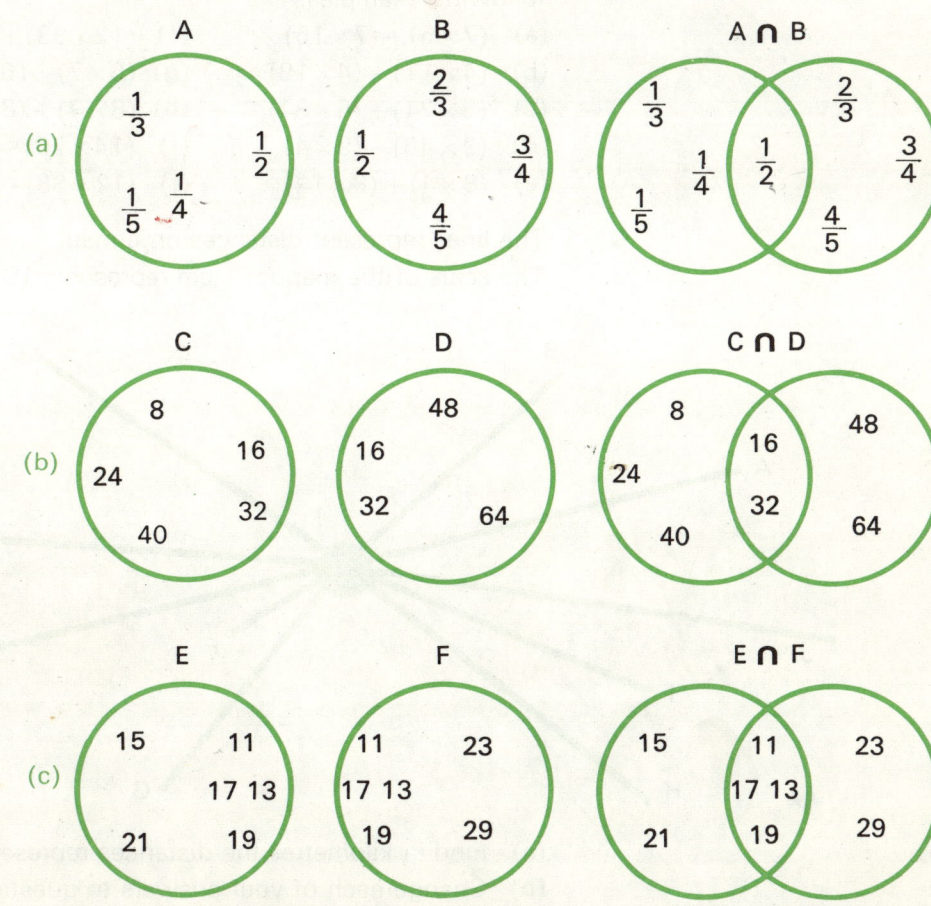

4 Write out the members of the following sets and intersections:

(a) A = {months beginning with the letter J}
B = {months with 31 days}
A ∩ B

(b) C = {vowels in the word 'facetious'}
D = {vowels in the alphabet}
C ∩ D

(c) E = {multiples of 10 between 9 and 51}
F = {multiples of 5 between 4 and 51}
E ∩ F

(d) G = {multiples of 3 between 20 and 40}
H = {multiples of 9 between 8 and 46}
G ∩ H

B 1 (a) 11×5 (d) 1×6 (g) $7 \times 1 \times 7$ (j) 100×100
(b) 5×11 (e) 7×0 (h) 4×5 (k) 300×300
(c) 6×1 (f) 0×7 (i) 4×50 (l) $2 \times 7 \times 25$

2 You may know an easy method for some of the following examples.
(a) 11×35 (e) 50×64 (i) 25×33 (m) 120×120
(b) 27×40 (f) 500×18 (j) 250×24 (n) 50×31
(c) 25×44 (g) 101×201 (k) 39×67 (o) 125×33
(d) 17×99 (h) 98×42 (l) 43×134 (p) 999×81

3 Use the method used in Exercise B, Chapter 2, to work out the
following examples.
(a) $(7 \times 5) + (7 \times 15)$ (f) $(12 \times 93) + (12 \times 7)$
(b) $(4 \times 11) + (4 \times 19)$ (g) $(6 \times 7) + (6 \times 5) + (6 \times 8)$
(c) $(5 \times 2\frac{1}{2}) + (5 \times 3\frac{1}{2})$ (h) $(8 \times \frac{3}{5}) + (8 \times \frac{2}{5})$
(d) $(9 \times 46) + (9 \times 4)$ (i) $(14 \times 20) + (14 \times 30) + (14 \times 50)$
(e) $(8 \times \frac{1}{4}) + (8 \times 1\frac{3}{4})$ (j) $(12 \times 38) - (12 \times 30)$

C 1 The lines represent distances on a map.
The scale of the map is: 1 cm represents 10 km.

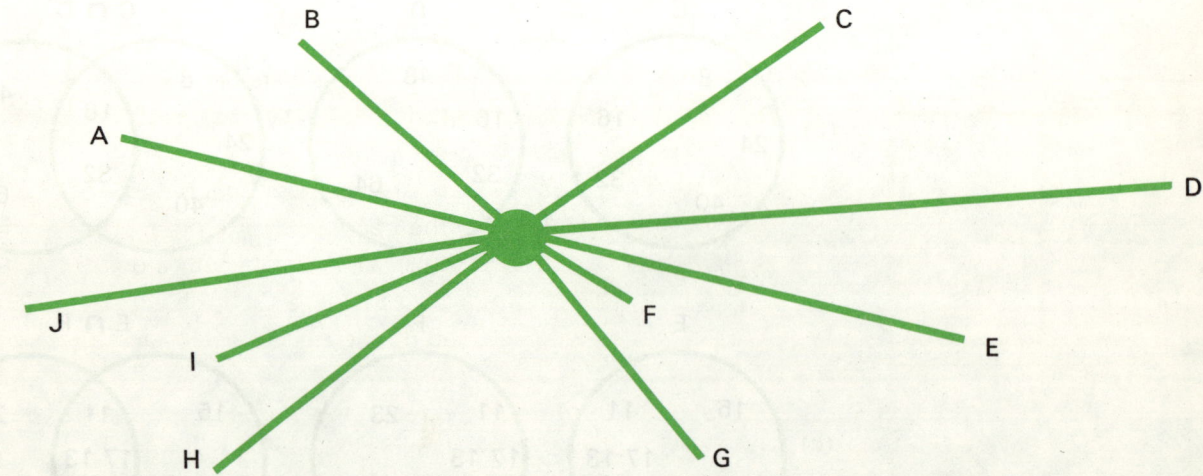

(a) Find in kilometres the distances represented by the lines.
(b) Change each of your answers to question (a) to metres.

2 Change to kilometres:

(a) 4,000 m	(e) 9,500 m	(i) 205 m	(m) 2,410 m
(b) 40,000 m	(f) 9,050 m	(j) 3,420 m	(n) 240,000 m
(c) 12,000 m	(g) 9,005 m	(k) 3,042 m	(o) 240,200 m
(d) 12,500 m	(h) 250 m	(l) 15,000 m	(p) 305,705 m

D 1

Halves	1						2				
Thirds	1			2				3			
Quarters	1			2		3			4		
Sixths	1		2		3		4		5		6
Twelfths	1	2	3	4	5	6	7	8	9	10	11	12

Use the above diagram to help you with the following examples.

(a) $\frac{1}{2}+\frac{1}{6}$

(b) $\frac{2}{3}+\frac{1}{6}$

(c) $\frac{1}{3}+\frac{5}{12}$

(d) $\frac{1}{2}+\frac{1}{4}+\frac{1}{12}$

(e) $\frac{1}{3}+\frac{1}{12}$

(f) $\frac{7}{12}+\frac{1}{4}$

(g) $\frac{1}{2}+\frac{7}{12}$

(h) $\frac{1}{2}+\frac{1}{4}+\frac{1}{12}$

(i) $\frac{1}{3}+\frac{1}{12}+\frac{1}{6}$

(j) $\frac{1}{6}+\frac{2}{3}+\frac{1}{4}$

(k) $1\frac{1}{12}+1\frac{1}{2}$

(l) $2\frac{2}{3}+2\frac{5}{12}$

(m) $1\frac{1}{4}+2\frac{1}{6}+3\frac{5}{12}$

(n) $5\frac{11}{12}+2\frac{1}{4}$

(o) $1\frac{3}{4}+2\frac{2}{3}$

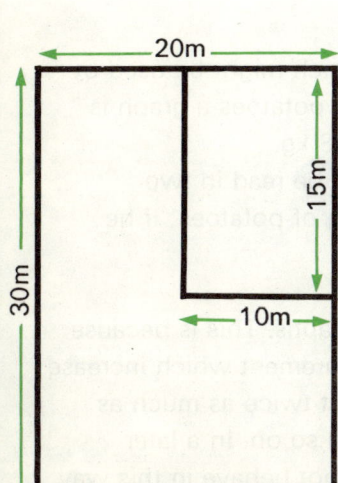

2 (a) Write down the set of natural numbers between 4 and 9.

(b) Make as many different proper fractions as possible using pairs of numbers from your set.

(c) Represent each fraction by a line on a graph.

(d) Use your graph to list the fractions in size order, smallest first.

3 (a) On a day's outing at the seaside Janet spent a quarter of her money in the morning and three eighths of her money in the afternoon. She then had 60p left. How much did she start off with?

(b) Five children out of a class of thirty are absent from school. What fraction of the class is present?

(c) Mr Bradley's garden is 30 m long by 20 m wide. In one corner he has a vegetable patch 15 m long by 10 m wide. What fraction is the vegetable patch of the whole garden?

(d) In the end of term examinations Julie obtained the following marks out of 20:

English	Maths.	Science	History	Geography
14	12	16	17	16

Find what fraction of the total possible marks she obtained, giving your answer in its simplest form.

Conversion Graphs

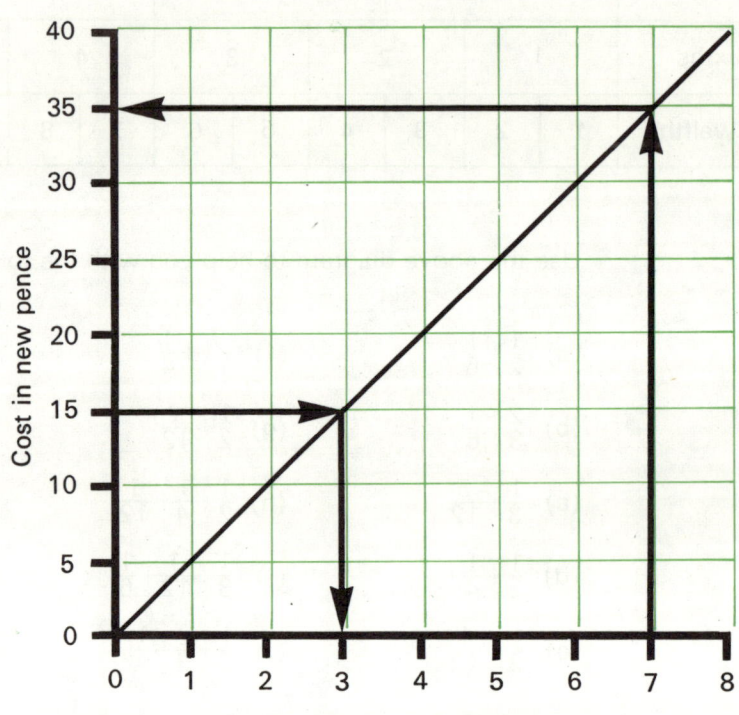

weight of potatoes in kilogrammes

The above graph is a kind of ready reckoner which might be used by a greengrocer. Having fixed the price of 1 kg of potatoes a graph is then drawn to show the cost of amounts up to 8 kg.

The coloured lines and arrows show how it can be read in two different ways. For 15p a customer will get 3 kg of potatoes; if he wishes to buy 7 kg the cost will be 35p.

This kind of graph is called a **conversion graph**.

All the graphs in this chapter are straight line graphs. This is because in each case we have chosen two sets of measurement which increase at the same rate. In the above example 2 kg cost twice as much as 1 kg, 3 kg cost three times as much as 1 kg and so on. In a later chapter we shall deal with examples which do not behave in this way. Can you think of any such example at the moment?

The diagrams explain the stages in preparing a graph to show:

(a) The number of kilometres a train travelled in a given number of hours.

(b) The number of hours the train took to travel a given number of kilometres.

1 Draw the axes at right angles **2 Mark in the scales**

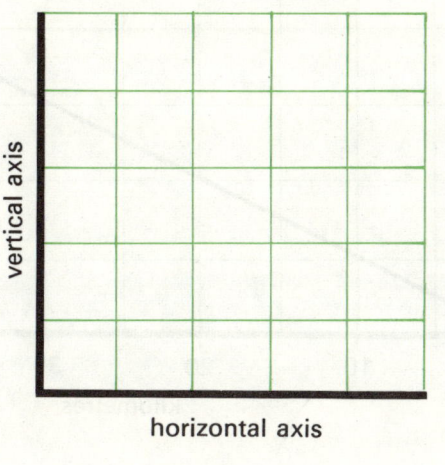

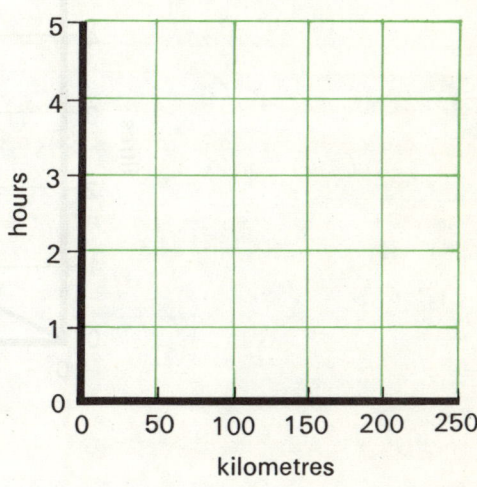

3 Plot the points **4 Draw the line through the points**

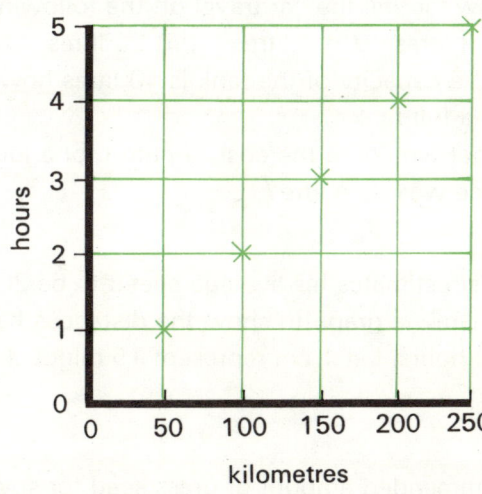

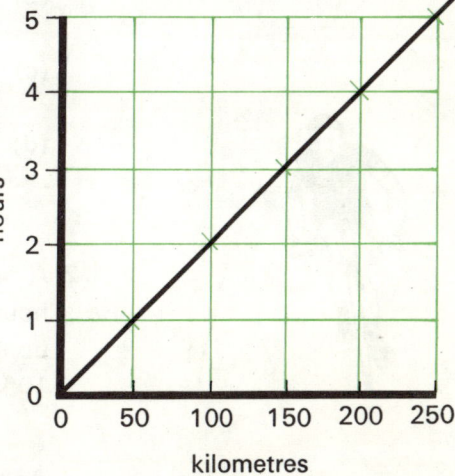

How many kilometres did the train travel in (a) 2 hours, (b) 5 hours?
How many hours did it take the train to travel (a) 50 km, (b) 150 km?
You will see that several points have been plotted. This has been done for two reasons; first because the practice is valuable, and second because each point acts as a check on the others. What is the least number of points needed to fix the direction of a particular straight line?

Exercises

A 1 The graph shows the connection between the distance a car travels and the amount of petrol it uses. Study it carefully and then answer the questions.

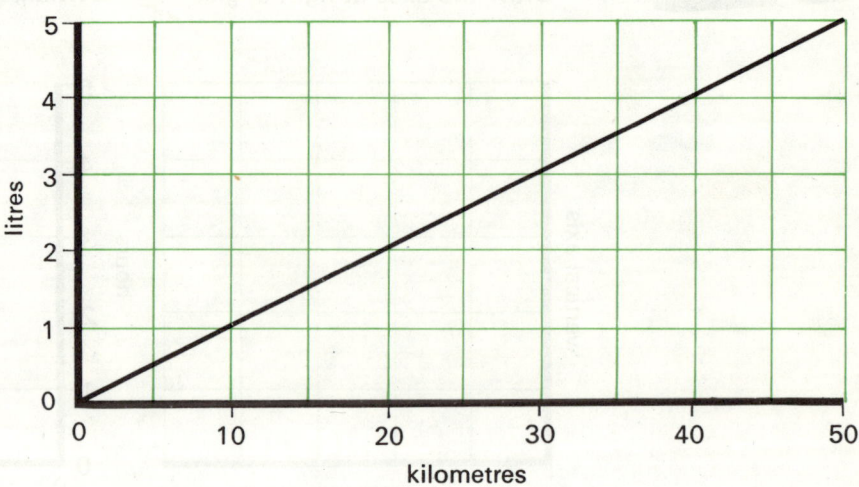

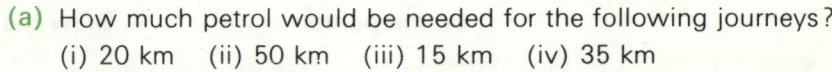

(a) How much petrol would be needed for the following journeys?
(i) 20 km   (ii) 50 km   (iii) 15 km   (iv) 35 km

(b) How far will the car travel on the following amounts of petrol?
(i) 2 litres   (ii) 4 litres   (iii) $2\frac{1}{2}$ litres   (iv) $1\frac{1}{2}$ litres

(c) If the capacity of the tank is 40 litres how far will the car travel on a full tank?

(d) What would be the cost of petrol for a journey of 100 km if the price was 8p a litre?

2 A cyclist estimates his average speed to be 20 km/h (kilometres per hour). Draw a graph to show the distances he would cover for times up to 3 hours. Let 1 cm represent 15 minutes and 1 cm represent 10 km.

3 A recommended amount of grass seed for sowing a lawn is 70 grammes to 1 square metre. Draw a graph to show the amount of seed needed for areas up to 500 square metres. Let 1 cm represent 10 g and 1 cm represent 50 square metres.

B 1 John prefers to save up for things costing a lot of money rather than spend all his pocket money each week. His conversion graph tells him how many weeks he will have to save for any particular item. Use the graph to answer the questions that follow.

SAVINGS

(a) How much money does John save each week?

(b) How many weeks will it take him to save:
   (i) 80p   (ii) £1·20   (iii) £1·80   (iv) 60p?

(c) How much can he save in:
   (i) 2 weeks   (ii) 8 weeks   (iii) 5 weeks   (iv) 7 weeks?

(d) He is given 50p for his birthday. If he has no other money to start with, how long will it take him to save enough money to buy a model aeroplane costing £2·30?

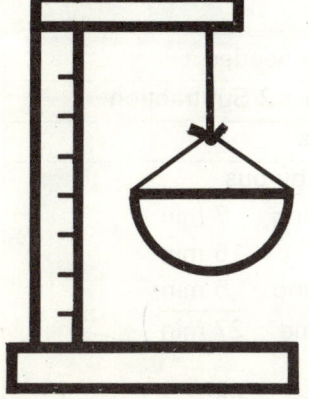

2   A simple spring balance can be made using a piece of wood, elastic and a tray. The table shows the length of a piece of elastic when various weights are put on the tray. Draw a graph to show these results.

Weight in grammes	0	10	20	30	40	50
Length of elastic in centimetres	20	28	36	44	52	60

3   During a thunderstorm the flash of the lightning and the bang of the thunder happen at the same moment; but light travels faster than sound and so we see the flash before we hear the bang.
For every 3 seconds between seeing the lightning and hearing the thunder the storm is approximately 1 km away. Draw a graph to show the distance between the observer and the storm for intervals of 3 seconds up to 30 seconds.

Addition and Subtraction

In earlier books we have learned how to add and subtract. It is not sufficient, however, to know **how** to do these operations, we must also know **when** to use them.

In the exercises you must decide whether to add or to subtract, or whether both operations are necessary.

First study carefully the following notes and example.

Addition (finding the sum or total)	Subtraction (finding the difference)
439 +276 ——— 715 11	424 −167 ——— 257

1 Always read the question at least twice.
2 Decide what kind, or kinds, of operation are needed.
3 Write out clear statements showing your working, step by step.
4 Always put in the units of measurement.

Example:

A boy usually cycles to school but goes by bus in bad weather. When he uses the bus he has a seven minute walk to the bus-stop, the bus journey takes a quarter of an hour, and he then has a five minute walk to school. It takes him half an hour cycling. Which way takes longer, and by how many minutes?

Operations needed		
1 Addition	**2** Subtraction	
Statements		
Travelling by bus		
Walking	7 min	
Bus	15 min	
Walking	5 min	
Total time	27 min	
Cycling		
Total time	30 min	
Difference in time		
= (30−27) min		
= 3 min		
Answer: Cycling takes 3 min longer than going by bus		

Exercises

A 1 When Sally checked up on her picture card album, she found she needed three cards on one page, two cards on another page, and four cards on a third page. The remaining pages were full. How many cards had she got out of a set of fifty?

2 On Monday morning Timothy's mother gave him a 50p piece to take to school. He needed 25p for the school bank, 12p for a school journey, 4p for bus fares and could have 3p to spend himself. How much change should he take home?

3 At half past seven Jane found she had forty minutes left to do her homework before watching her favourite television programme; after that, she had to go to bed. If the programme lasted thirty-five minutes, at what time did she go to bed?

4 Mr Burton decided to buy some new carpet. On measuring he found he needed 4 m 50 cm for the hall, 5 m 40 cm for the staircase, and 2 m 25 cm for the landing. What was the total length of carpet he required? If the carpet was sold only in exact metre lengths, what length must he buy and how many centimetres would be spare?

5 On checking her weight Mary found that she weighed 81 kg 50 g. She worked out that she weighed 65 g more than she weighed a month earlier. What was her previous weight?

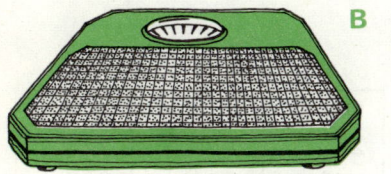

B 1 The plan shows the routes taken for the senior and junior cross country runs. By measuring, find the total distance in each case, and how much further the seniors had to run than the juniors.

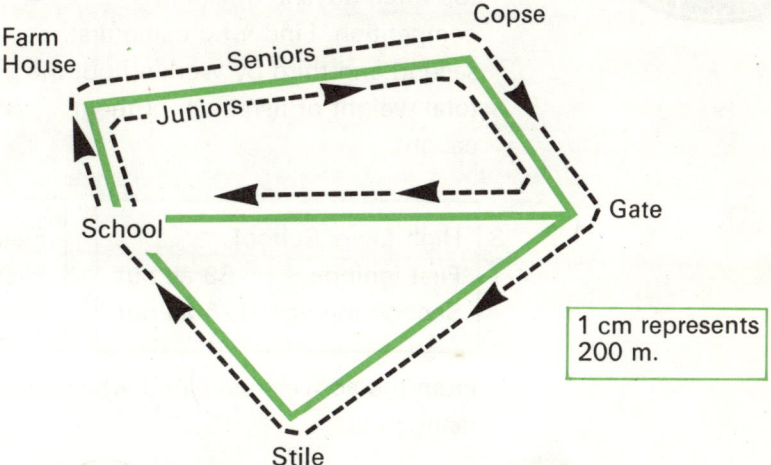

1 cm represents 200 m.

2 A prefect was given the task of finding the time taken for fire drill. The bell sounded at one and a half minutes to three, and the checking of the last form was completed at three minutes twenty seconds after three. What was the total time in minutes and seconds?

3 From the table find:
(a) The term totals and position for each team.
(b) The difference between the totals for the first and last teams.

Form 4 North Team points for Summer Term			
Reds	Yellows	Blues	Greens
May	35	41	38	48
June	42	37	34	43
July	29	24	17	35

4 The list shows the matches to be played against another school. Two 40 seater buses were ordered to take the players and four teachers. How many seats were left for spectators?

	Team	Reserves
Rugby	15	2
Soccer	11	2
Hockey	11	1
Netball	7	1

5 A school hall could seat 500 people. At a school concert 25 reserved seats had been occupied by special visitors, not needing tickets. 257 adult tickets and 169 children's tickets had been collected. How many seats were empty?

C **1** John, Harry and Tony are playing a game of darts in which they have to score exactly 301 to win.
Find how many each of them still needs after his first three throws.

	John	Harry	Tony
1st Dart	7	5	double 8
2nd Dart	19	treble 6	15
3rd Dart	50	17	16

2 Dennis, Robert and Derek were the top three boys in a fishing competition. Find who came first, second and third by working out the total weight of fish each of them caught.

Dennis	Robert	Derek
240 g	380 g	275 g
145 g	240 g	230 g
120 g	365 g	185 g
175 g		295 g
160 g		

3

High Ellers School	
First innings:	89 all out
Second innings:	125 all out

Fieldwood School	
First innings:	158 for 7 declared
Second innings:	73 all out

From the above scores find which team won the match, and by how many runs.

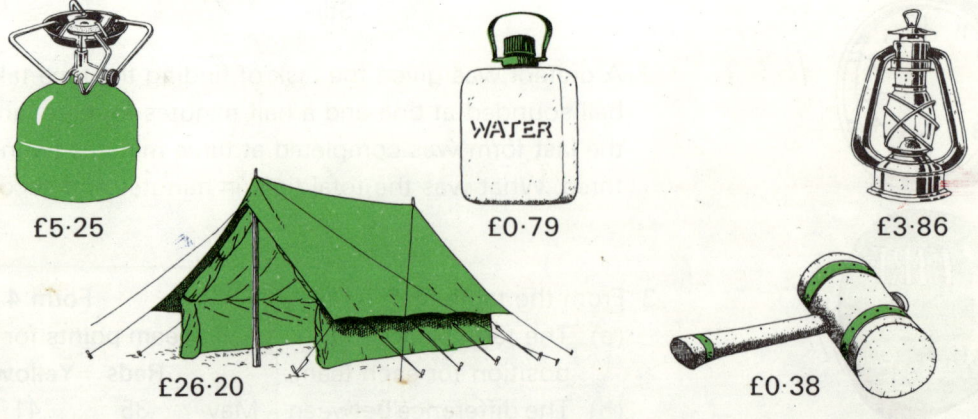

£5·25   £0·79   £3·86

£26·20   £0·38

4 Find the total cost of the camping equipment shown above.
If a school received from the parents a grant of £27·50 towards the cost, how much would the school have to pay?

5 In a long jump competition Keith jumped 5m 25 cm, Ian 4 m 95 cm and Paul 5 m 40 cm. Find the average distance jumped. How much further did Paul jump than Ian?

Volume and Weight

Specific Gravity

In the Metric System the measurements of volume, weight and capacity are all connected.

1 gramme was taken to be the weight of 1 cubic centimetre of water.

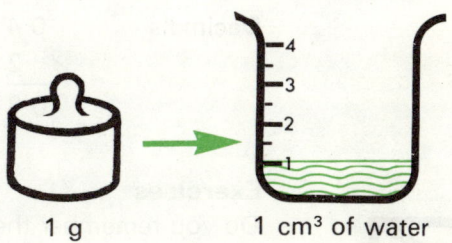

1 g   1 cm³ of water

To engineers, scientists and many other people in industry the weight of a substance or liquid is very important. Compare, for example, the weight of metal used for making aircraft with that used for making anchors.

The weights of substances and liquids are compared to the weight of an equal volume of water, thus

1 cm³ of water weighs 1 g, 1 cm³ of lead weighs 11·4 g

so lead is nearly eleven and a half times as heavy as water.

We say **the Specific Gravity of water is 1**,

so **the Specific Gravity of lead is 11·4.**

Since petrol is lighter than water its specific gravity, 0·75, is less than 1. What happens if petrol is poured on water?

Multiplying Decimals

We have already learned the methods for adding and subtracting decimal fractions. We now need to know how to multiply decimals by whole numbers.

You will remember that a decimal point fixes the place of a figure in a number and so gives it a particular value. When we are multiplying, therefore, we must put in the decimal point when we come to it, so that each figure remains in its correct column. The examples will make it clear.

Whole numbers	4	12	25	137
	$\times 2$	$\times 3$	$\times 5$	$\times 4$
	8	36	125	548

Decimals	0·4	1·2	2·5	1·37
	$\times$ 2	$\times$ 3	$\times$ 5	$\times$ 4
	0·8	3·6	12·5	5·48

Exercises

Do you remember the following facts about volume?

(i) When we find the volume of a container we are measuring the space inside it.

(ii) The unit of measurement for volume is a cube.

(iii) To find the volume of a rectangular solid, a cuboid, we multiply the length by the breadth by the height.

Example: Find the volume of a box 16 cm long, 10 cm wide and 4·5 cm deep.

$V = 1 \times b \times h$
$= (16 \times 10 \times 4·5) \text{ cm}^3$
$= 720 \text{ cm}^3$

A Now find the volumes of these containers, choosing the most suitable units in each case, that is, mm³, cm³ or m³.

1

2

3
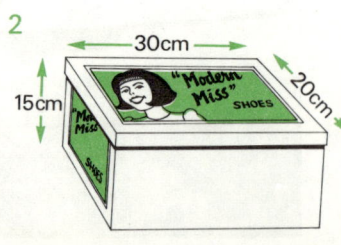

4
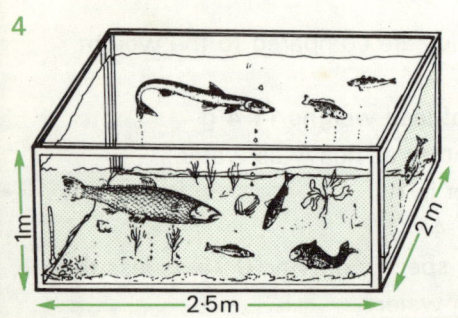

5
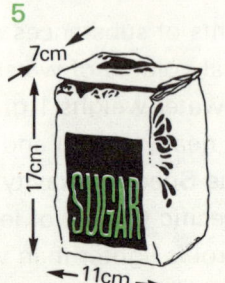

6
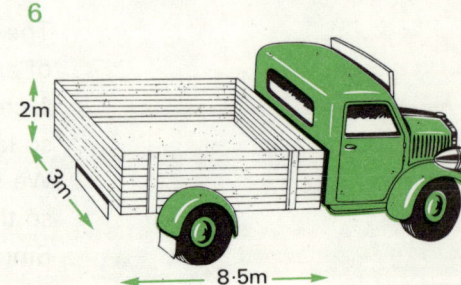

B $A = \{0\cdot4, 0\cdot8, 1\cdot2, 1\cdot6, 2\cdot0, 2\cdot4, 2\cdot8, 3\cdot2, 3\cdot6\}$

1 Multiply each member of set A by 10 and then make each of your answers a member of set B.

2 Write a rule to describe set B.

3 Multiply each member of set A by 2 and then make each of your answers a member of set C.

4 Multiply each member of set C by 5 and make each answer a member of set D.

5 Which of the sets A, B, C and D are equal. Can you say why?

The table shows the Specific Gravity of some common substances. You will need to refer to it to answer the questions in sections C and D.

Specific Gravity	
Petrol	0·75
Ice	0·9
Water	1·0
Aluminium	2·7
Iron	7·86
Copper	8·9
Lead	11·4
Gold	19·3
Platinum	21·5

Notice that

> 1 cm³ of water weighs 1 g
>
> therefore   1 cm³ of aluminium weighs 2·7 g
>
> 100 cm³ of aluminium weigh (100 × 2·7) g
>
> i.e. 270 g

C   Very often it is useful to obtain a rough answer to a question before attempting an exact calculation. We shall see later, in fact, that there is no exact answer to many problems, and we shall have to decide how accurate we need to be. For the questions in this exercise give an answer to the nearest whole number.

About how many times is:

1 gold heavier than water?   5 platinum heavier than lead?

2 copper heavier than water?   6 lead heavier than aluminium?

3 gold heavier than copper?   7 platinum heavier than aluminium?

4 iron heavier than aluminium?   8 aluminium heavier than ice?

D 1 Find in kilogrammes the weight of 3,500 cm³ of water.

2 Find the weight of a bar of iron 100 cm long, 5 cm wide and 2 cm thick.

3 Find the weight of a 10 cm cube of lead.

4 Find the difference between the weight of 1 cm³ of lead and 1 cm³ of copper.

5 How much heavier is 1,000 cm³ of water than 1,000 cm³ of petrol?

6 Which is the heaviest metal in the table? Which is the lightest metal in the table? What is the difference between weights of 1 cm³ of each of these metals?

7 (a) What is the weight of 1,000 cm³ of petrol? (b) What is the weight of 9,000 cm³ of petrol? (c) What is the weight of 9 litres of petrol?

8 Find the weight of a bar of gold 50 cm long, 20 cm wide and 10 cm thick.

Approximation

Can you measure accurately?

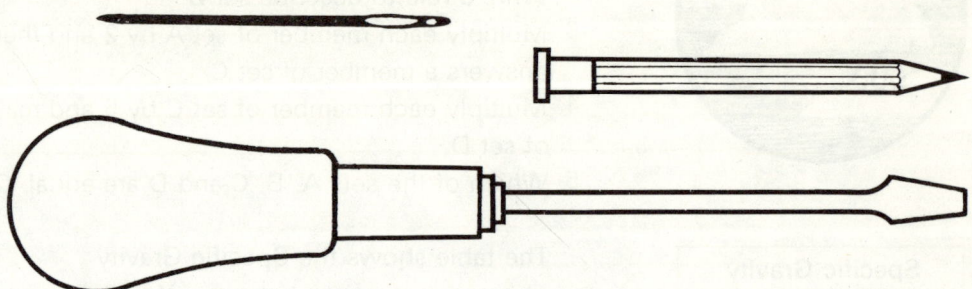

Measure the lengths of the three objects above and then compare your answers. The chances are that you do not all agree; can you suggest why?

If your answer is correct to the nearest millimetre you have been reasonably accurate, but what about an answer to the nearest **tenth** of a millimetre, or the nearest **hundredth** of a millimetre?

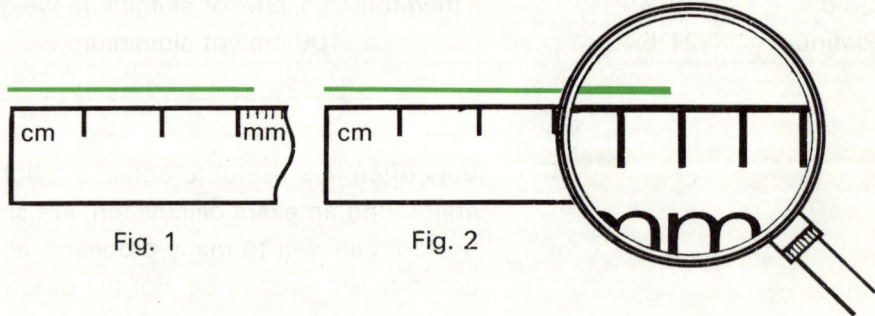

Fig. 1   Fig. 2

Fig. 1 shows a line which appears to be about 3 cm 2 mm long.
Fig. 2 shows another line of the same length but with the final millimetre as it might be seen through a magnifying glass. We can now see that the line is about 3 cm 1·8 mm long.
Perhaps you have realized that **it is impossible to measure exactly**. We can only be as accurate as our observation and measuring instruments allow.

How accurate do we need to be?

Look at the following pairs of statements and decide which of each pair is the more sensible.

The car was travelling at 31·375 km an hour.

The car was travelling at about 30 km an hour.

The distance from Doncaster to London is 263·59 km.

The distance from Doncaster to London is about 260 km.

This packet contains 253 grammes.

This packet contains approximately 250 grammes.

You will see that the statements are easier to understand when we use a simpler number which is about the correct value. In this case we are using an **approximate** measurement, which often is all we require. The following diagram shows how we can find an approximate value for the length of a line by measuring it to the nearest centimetre.

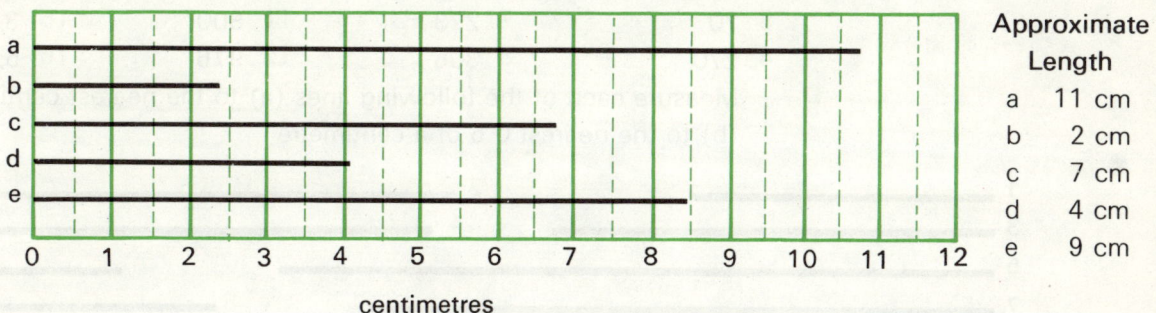

If the fraction of a centimetre is less than a half we take the previous whole number, if the fraction is more than a half we take the next whole number. Line 'e' is 8·5 cm, which is exactly halfway between 8 cm and 9 cm, and in this case it is usual to take the higher value.

Exercises
A

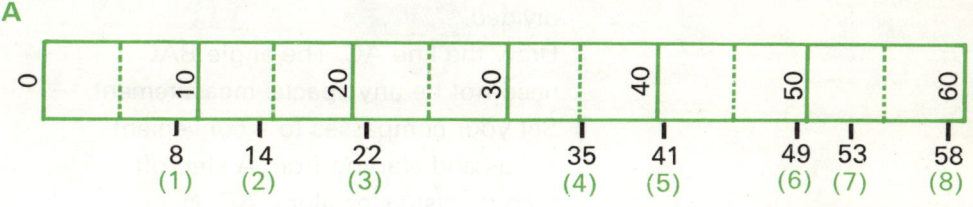

For each of the examples, (1) to (8), find an approximate value which is correct to the nearest 10.

B

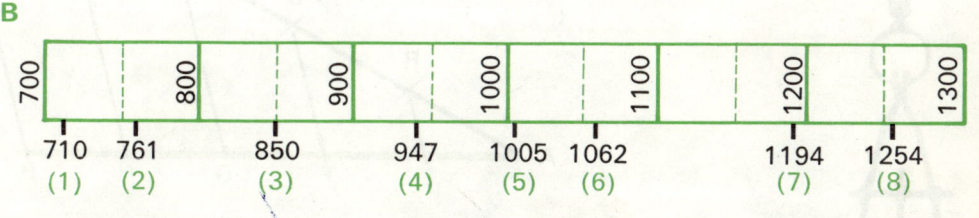

For each of the examples (1) to (8),

(a) find an approximate value which is correct to the nearest 100,

(b) find an approximate value which is correct to the nearest 10.

| 0 | 25 | 50 | 75 | 100 |

C For each of the following numbers find an approximate value which is correct to the nearest 50. Use the scale above to help you.

1	26	5	177	9	503	13	1,040
2	126	6	377	10	732	14	2,209
3	70	7	223	11	800	15	3,029
4	270	8	305	12	976	16	5,575

D Measure each of the following lines (a) to the nearest centimetre, (b) to the nearest 0·5 of a centimetre.

E Since it is difficult to measure accurately we avoid measuring whenever we can. The notes and diagram explain how to divide a line into 5 equal parts without measuring.

1 Draw a line AB any convenient length. This is the line to be divided.

2 Draw the line AC. The angle BAC need not be any special measurement.

3 Set your compasses to a convenient radius and starting from A step off 5 equal distances along AC. Name these points R, S, T, U and V.

4 Draw a line from V to B.

5 Using a ruler and setsquare draw UH, TG, SF and RE parallel to VB.

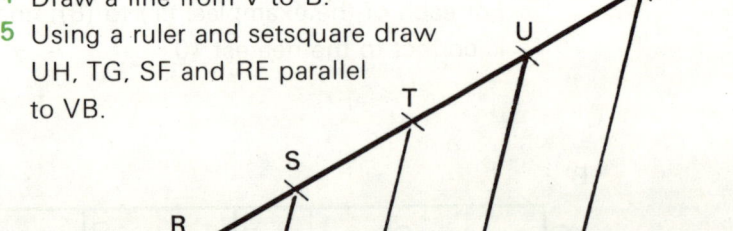
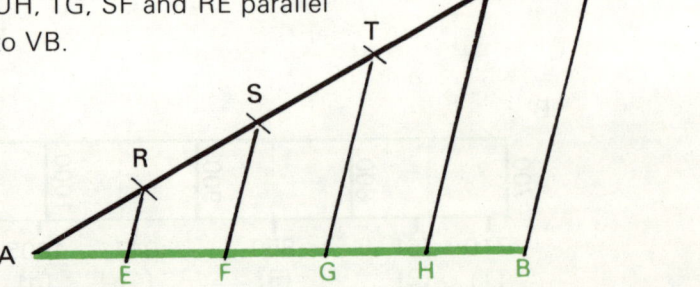
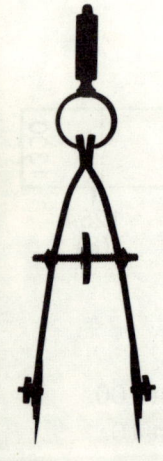

When you have studied the above method, draw three lines and divide them into (a) 6 equal parts, (b) 7 equal parts, (c) 10 equal parts.

Explain why the best method of checking your work is by using a pair of dividers.

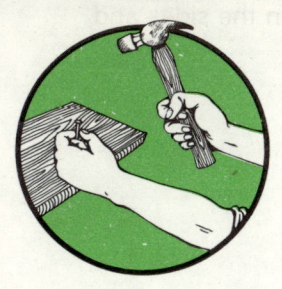

F Rewrite these sentences, replacing each number by a simpler number which is easier to understand. You are told in the brackets how close your answer should be.

1 John's garden is **37·2 m** long by **19·8 m** wide. (nearest unit)
2 The recommended tyre pressure is **1·69 kg** per square centimetre. (nearest tenth)
3 The distance from London to Leeds is about **303 km**. (nearest ten)
4 The distance from the Earth to the Moon is **384,317 km**. (nearest thousand)
5 Alan is using a hammer weighing **1,980 g**. (nearest 100)
6 The newspaper has a circulation of about **205,560**. (nearest thousand)
7 The car has a top speed of about **235 km** per hour. (nearest ten)
8 Janet is **153·4 cm** tall and weighs **43·6 kg**. (nearest unit)
9 The cost of the repairs was about **£973·50**. (nearest £50)
10 The petrol tank holds **40·55 litres**. (nearest unit)

Making Sure—Two

A 1 (a) Draw a graph similar to the one shown, making each axis 10 cm long.
(b) Mark in the points (0, 0), (2, 1), 4, 2).
(c) Draw a straight line through these points right across your graph.
(d) Mark in 5 more points on your line and say what the ordered pair is for each one.
(e) Say what you notice about the numbers in each of the ordered pairs on the line.

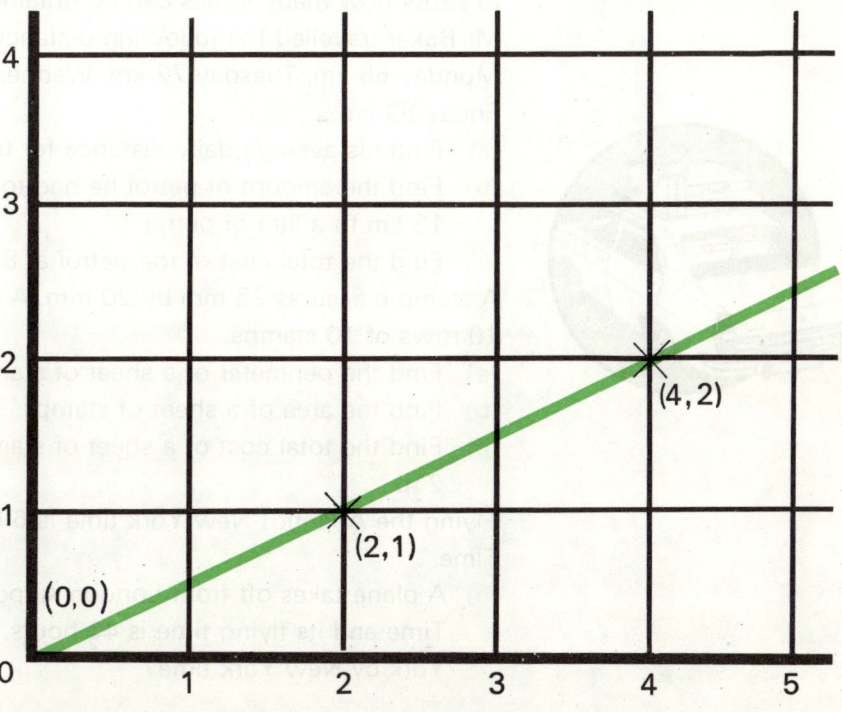

2 The following graph shows the connection between the sides and perimeters of squares.

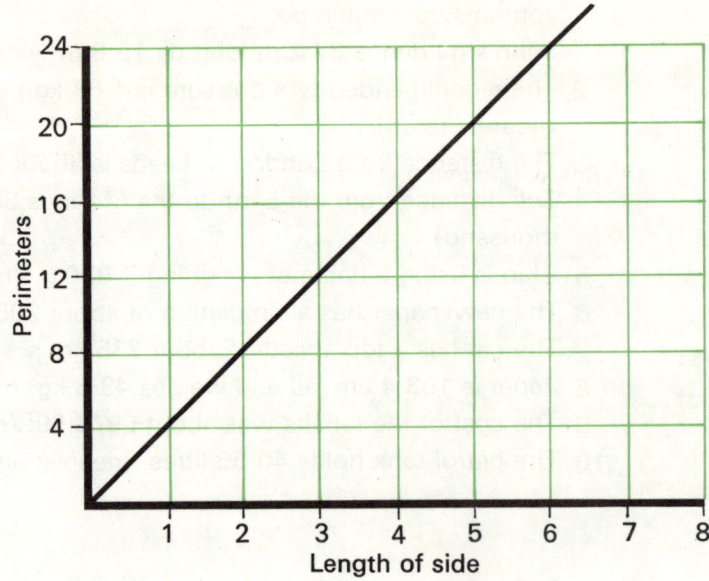

Notice the difference in the scales for the two axes and that no units of measurement have been given. Can you say why?

(a) Find from the graph the perimeters of squares whose sides measure:

  (i) 3 cm   (ii) 5 cm   (iii) $2\frac{1}{2}$ m   (iv) $4\frac{1}{2}$ mm

(b) Find the length of the sides of squares whose perimeters measure:

  (i) 16 m   (ii) 24 mm   (iii) 22 cm   (iv) 2 m

(c) Draw a graph to show the connection between the perimeters and sides of equilateral triangles.

3 If £1 can be exchanged for 13 French francs, draw a conversion graph to show how many francs can be obtained for amounts up to £10.

B 1 Mr Baker travelled the following distances on a touring holiday: Monday 65 km, Tuesday 72 km, Wednesday 51 km, Thursday 54 km, Friday 83 km.

(a) Find his average daily distance for the tour.

(b) Find the amount of petrol he had to buy if the car averaged 13 km to a litre of petrol.

(c) Find the total cost of the petrol at 8p a litre.

2 A stamp measures 25 mm by 20 mm. A sheet of stamps is arranged in 10 rows of 10 stamps.

(a) Find the perimeter of a sheet of stamps.

(b) Find the area of a sheet of stamps.

(c) Find the total cost of a sheet of stamps if each stamp is valued at $2\frac{1}{2}$p.

3 Flying the Atlantic! New York time is 6 hours behind British Summer Time.

(a) A plane takes off from London Airport at 09.45 British Summer Time and its flying time is $4\frac{1}{2}$ hours. What time will it land in New York by New York time?

(b) A plane takes off from New York at 15.40 and its flying time is $3\frac{3}{4}$ hours. What time will it land at London Airport by British Summer Time?

C 1 Find the volume of each of these containers:

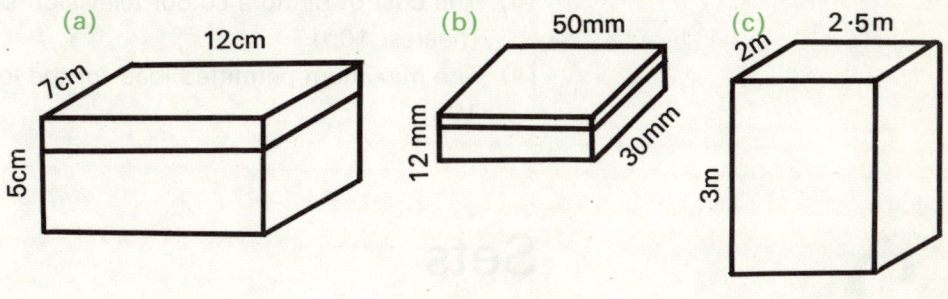

2 (a) 10×12
    (b) 10×1·2
    (c) 20×1·2
    (d) 9×19
    (e) 9×1·9
    (f) 11×2·4

    (g) 6×15 cm
    (h) 6×1·5 cm
    (i) 4×69 cm
    (j) 4×0·69 cm
    (k) 7×1·08 cm
    (l) 8×2·37 cm

    (m) 5×£28
    (n) 5×£2.80
    (o) 7×£0.85
    (p) 12×£2.07
    (q) 3×£41.20
    (r) 30×£41.40

3 For these examples you will need to refer back to the Specific Gravity table on page 29.
    (a) Find the weight of 15 cm³ of aluminium.
    (b) Find the weight of a 10 centimetre cube of ice.
    (c) Find the difference in weight between 20 cm³ of water and 20 cm³ of ice.
    (d) Find the weight of 40 litres of petrol (1 litre = 1,000 cm³).

D 1 Use a ruler, compasses and setsquare to divide a line into 8 equal parts.

2 For each of the following numbers find an approximate value which is correct **(i)** to the nearest 10, **(ii)** to the nearest 100, **(iii)** to the nearest 1,000.

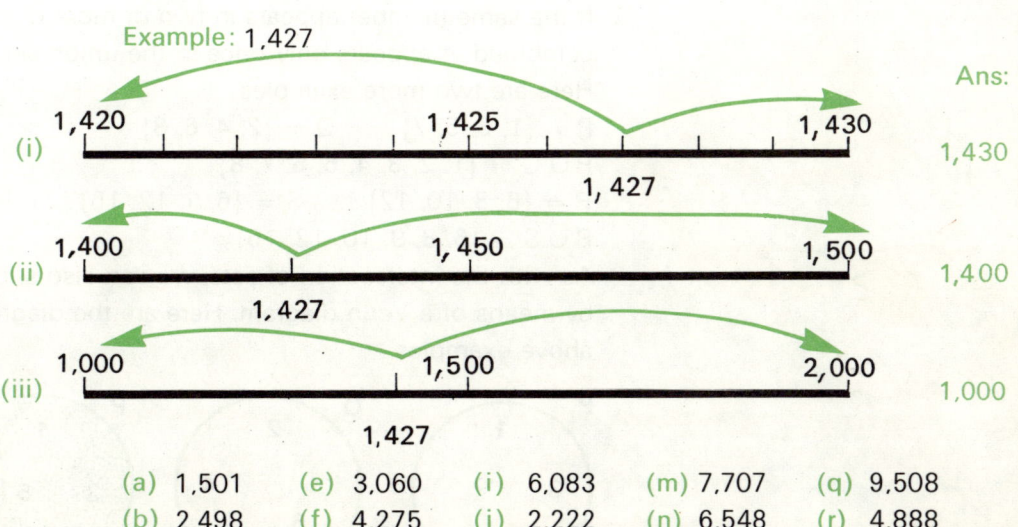

Example: 1,427

    Ans:
    (i)   1,430
    (ii)   1,400
    (iii) 1,000

(a) 1,501	(e) 3,060	(i) 6,083	(m) 7,707	(q) 9,508
(b) 2,498	(f) 4,275	(j) 2,222	(n) 6,548	(r) 4,888
(c) 2,989	(g) 874	(k) 5,555	(o) 6,553	(s) 10,009
(d) 3,006	(h) 5,086	(l) 7,077	(p) 9,432	(t) 11,694

3 Rewrite these sentences replacing each number by an approximate value.

(a) The desk is 150·7 cm long by 92·3 cm wide. (nearest unit)

(b) The bottle holds 746 ml. (nearest 10)

(c) There are approximately 973 pupils on roll. (nearest 100)

(d) The cost of hiring a colour television set is about £1·46 a week. (nearest 10p)

(e) The maximum permitted load for the lorry is 1,030 kg. (nearest 100)

Sets

Do you remember what these signs mean? $\varnothing$ $\cap$

In this chapter we are going to investigate another of the signs used in set language.

When two or more sets are combined we call this operation the **union** of the sets, and we use this sign: $\cup$

For the combination of sets A and B

we say: **A union B**   and we write: $A \cup B$

Look carefully at this example and see if you can discover for yourselves how to make the union of two sets.

$A = \{\triangle \ \square \ \bigcirc\}$ $B = \{\bigcirc \ \triangle \ \hexagon\}$

$A \cup B = \{\triangle \square \bigcirc \triangle \hexagon\}$

There are two important things to know about the union of sets:

1 The union set includes **all** the members of the sets which are combined.

2 If the same member appears in two or more of the sets which are combined, it appears only **once** in the union set.

Here are two more examples:

P = {1, 3, 5, 7}   Q = {2, 4, 6, 8}

P $\cup$ Q = {1, 2, 3, 4, 5, 6, 7, 8}

R = {6, 8, 10, 12}   S = {6, 9, 12, 15}

R $\cup$ S = {6, 8, 9, 10, 12, 15}

As with the intersection of sets, we can also show the union of sets by means of a Venn diagram. Here are the diagrams to represent the above examples.

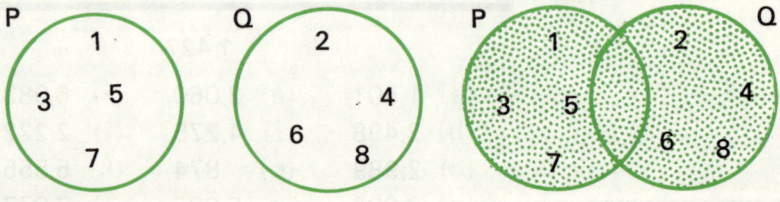

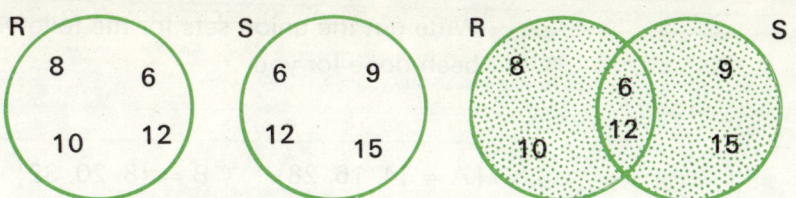

The colour has been added to show that all the members of both sets are included in the union set. Notice, especially, the position of the members which appear in both set R and set S.

Exercises

A   Write out the members of the following sets:

1 {multiples of 7 between 6 and 50}
2 {prime numbers between 40 and 70}
3 {even numbers between 989 and 1,005}
4 {multiples of 9 between 82 and 89}
5 {square numbers between 15 and 122}
6 {triangular numbers between 0 and 29}
7 {numbers between 19 and 81 which are exactly divisible by 4 and by 5}
8 {multiples of 20 between 950 and 1,050}

B   For the following examples write out the separate sets and the union sets represented by the diagrams.

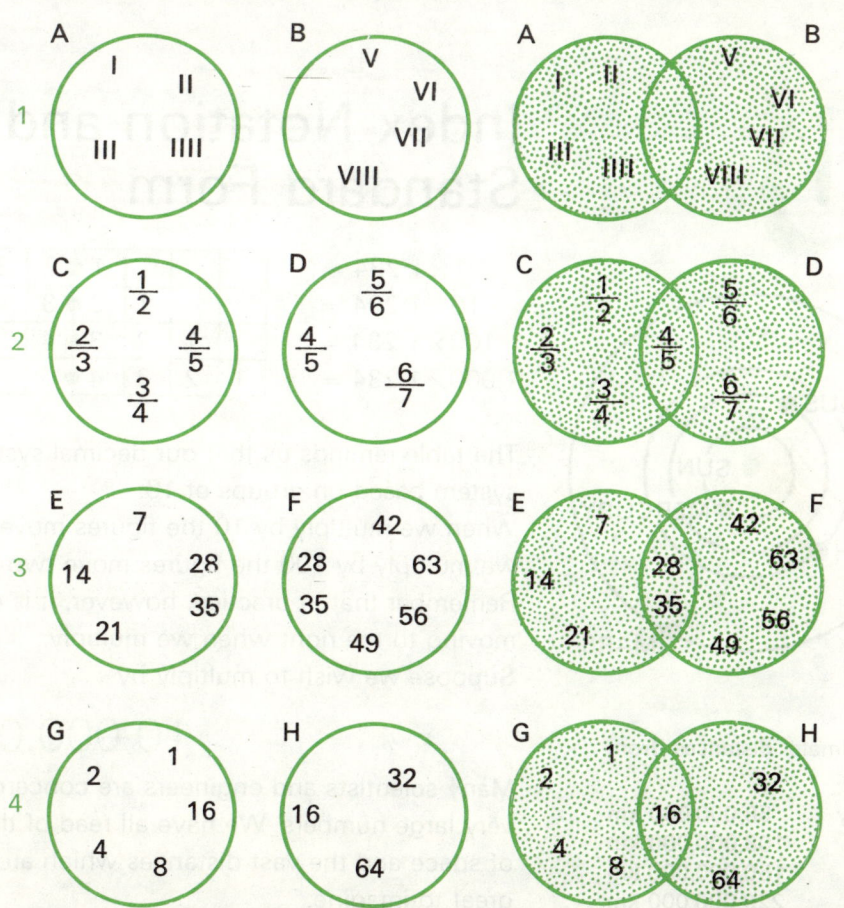

C   Write out the union sets for the following examples. The first one has been done for you.

> **1**   A = {4, 16, 28}   B = {8, 20, 32}   C = {12, 24, 36}
> A ∪ B ∪ C = {4, 8, 12, 16, 20, 24, 28, 32, 36}

2   E = { △ ○ □ }   F = { ○ □ ▱ ⬡ }

3   X = {1, 2, 3, 4}   Y = {4, 5, 6, 7}
4   P = {2, 4, 6, 8}   Q = {10, 12, 14, 16}
5   M = {36, 49, 64, 81}   N = {64, 81, 100, 121}
6   R = {c, h, e, a, p}   S = {p, e, a, c, h}
7   F = {3, 9, 15, 21}   G = {6, 12, 18, 24}
8   J = {48, 60}   K = {12, 24, 36}   L = {72, 84, 96, 108}
9   V = {+, −}   W = {×, ÷}   X = {=, >, <}
10   Y = {7, 77, 777}   Z = {7, 77, 777}

D   Write a rule to describe the union sets you have listed for each of the examples in Exercise C.
No. 1 will be: {multiples of 4 between 3 and 37}.

10

Index Notation and Standard Form

			1	2	3	4
		1	2	3	4	
	1	2	3	4		
1	2	3	4			

$1 \times 1 \cdot 234 =$
$10 \times 1 \cdot 234 =$
$100 \times 1 \cdot 234 =$
$1,000 \times 1 \cdot 234 =$

The table reminds us that our decimal system of counting is a 'place' system based on groups of 10.
When we multiply by 10 the figures move one place to the left, when we multiply by 100 the figures move two places to the left, and so on.
Remember that in practice, however, it is easier to think of the point moving to the right when we multiply.
Suppose we wish to multiply by

10,000,000

Many scientists and engineers are concerned with very large numbers. We have all read of the exploration of space and the vast distances which are almost too great to imagine.

Approximate distances from the Sun:

Mercury	58,000,000 km
Venus	108,000,000 km
Earth	150,000,000 km
Mars	229,000,000 km

To make it easier to work with these large numbers mathematicians have devised a kind of shorthand. The following table shows how it works.

Number	Powers of 10	Shorthand	Read as:
100	10×10	10^2	10 to the power 2 (10 squared)
1,000	$10 \times 10 \times 10$	10^3	10 to the power 3 (10 cubed)
10,000	$10 \times 10 \times 10 \times 10$	10^4	10 to the power 4
100,000	$10 \times 10 \times 10 \times 10 \times 10$	10^5	10 to the power 5

$$\text{base} \rightarrow 10^{2} \leftarrow \text{index}$$

When we write a number in this way we say we are using **Index Notation**.

Now let us see how we use index notation for writing big numbers. Study these three examples and try to work out for yourselves how it is done.

$$300,000 = 3 \cdot 00000 \times 100,000$$
$$= 3 \times 10^5$$

$$7,500,000 = 7 \cdot 500000 \times 1,000,000$$
$$= 7 \cdot 5 \times 10^6$$

$$125,000,000 = 1 \cdot 25000000 \times 100,000,000$$
$$= 1 \cdot 25 \times 10^8$$

The expressions underlined in colour are numbers written in **Standard Form.**

When a number is written in standard form it is expressed as a **number between one and ten** multiplied by a **power of ten**.

Here are three more examples:

$25,000 = 2 \cdot 5 \times 10^4$ (move the point 4 places to the left, multiply by 10^4)

$731,000 = 7 \cdot 31 \times 10^5$ (move the point 5 places to the left, multiply by 10^5)

$600,000,000 = 6 \times 10^8$ (move the point 8 places to the left, multiply by 10^8)

A Write down the following numbers then (i) underline the noughts which alter the place and value of any of the other digits, (ii) cross out those noughts which do not make any difference to the number.

Example: 4,0̲0̲8·5∅∅∅

1	0321	6	0000·509	11	0303·0303
2	3,021	7	12,000·300	12	250,000·009
3	6·400	8	090·090	13	000·7000
4	6·004	9	00017·70100	14	11,000,000·0008
5	000·07	10	302,408·0300	15	000500·008200

B Write the following numbers in words:

1	101	5	25·25	9	00750·50700
2	10·1	6	2,704·60	10	3,097·75
3	0·07	7	200,300·0090	11	0·075
4	0·009	8	1,030,708·04	12	10,500·037

C Rewrite this set of numbers expressing each member as a power of 10:
{100, 1,000, 10,000, 100,000, 1,000,000, 10,000,000, 100,000,000}

D Find the areas of these figures and then express your answers as powers of 10.

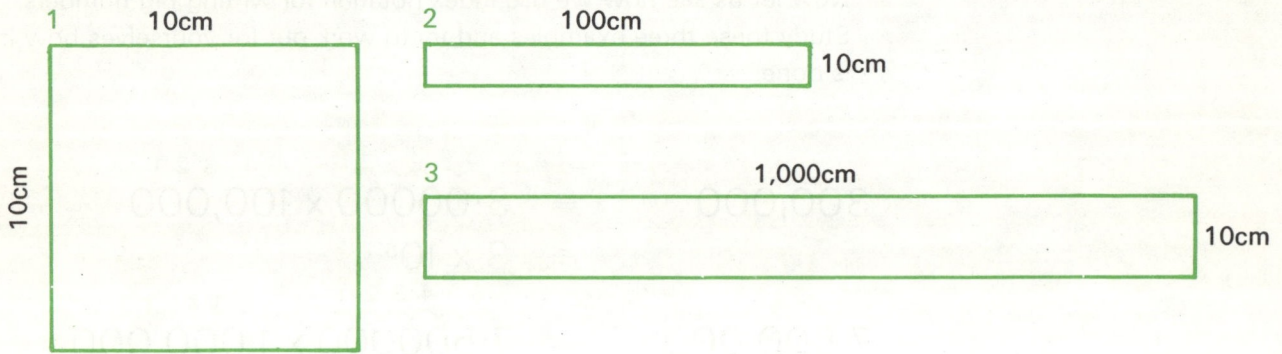

E Find the volumes of these rectangular prisms and then express your answers as powers of 10.

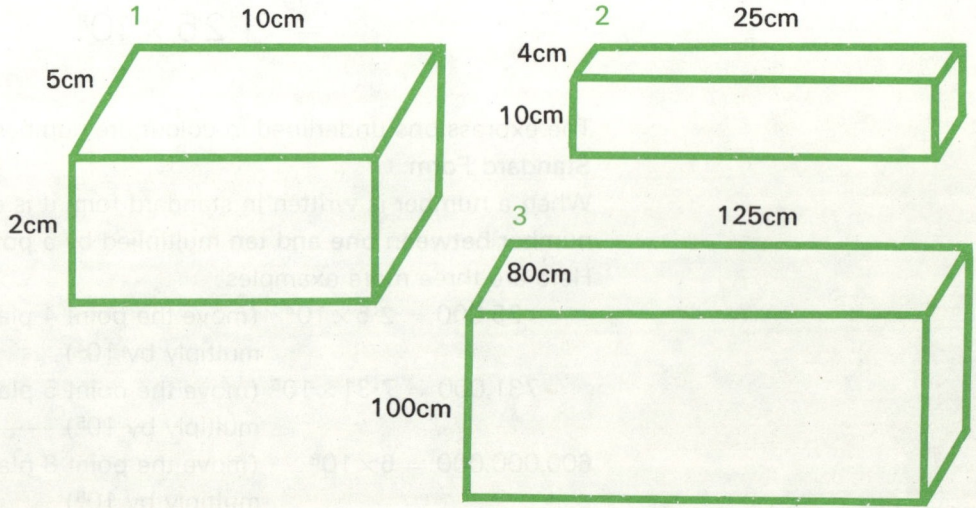

F Work out the following, giving each answer as an ordinary number.

1	10×35	6	$1{,}000 \times 3.45$	11	$10{,}000 \times 5$	16	$10 \times 10 \times 0.08$
2	10×3.5	7	$10{,}000 \times 0.001$	12	$10^4 \times 0.5$	17	$10^2 \times 2 \times 0.5$
3	100×12.3	8	$10^4 \times 0.003$	13	$10^3 \times 4.03$	18	$10^5 \times 123.6$
4	$10^2 \times 12.3$	9	$10^2 \times 37$	14	$10^5 \times 21.74$	19	$10^4 \times 40.007$
5	$10^3 \times 0.417$	10	$10^2 \times 0.037$	15	$10^6 \times 0.848$	20	$10^2 \times 10^3$

G Write out the numbers in the following sentences in standard form.

1 Electron microscopes can magnify more than **100,000 times.**

A fibre of wool magnified 1,200 times by an electron microscope.

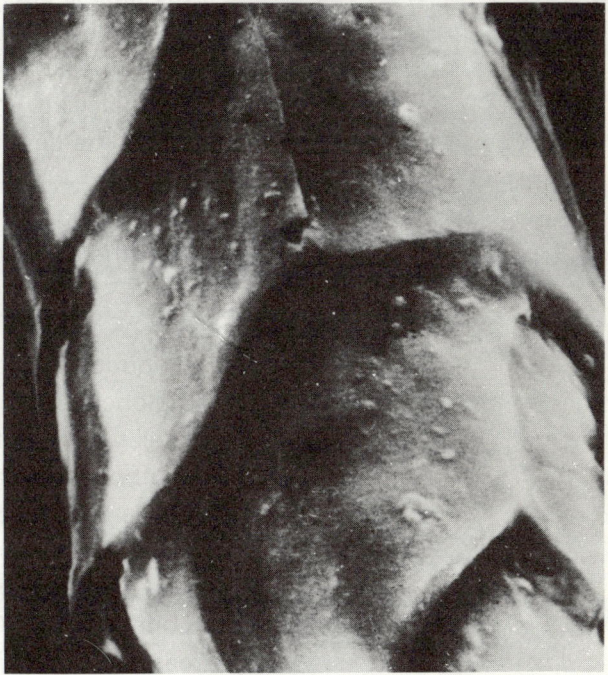

Here is the same fibre magnified 4,200 times. A postage stamp magnified to this extent would cover two football pitches!

2 The diameter of Mars is about **6,720 km.**

3 Light travels at a speed of about **298,000 km** per second.

4 The average distance between the earth and the moon is approximately **384,000 km.**

5 Some North Sea gas is found in layers of sandstone laid down over **250,000,000 years** ago.

6 The sun is about **150,000,000 km** from the earth, and its light takes about 8 minutes to reach us.

7 About **$2\frac{1}{2}$ million** slot machines had to be changed because of the introduction of decimal coinage.

8 The population of Turkey is about **24 million.**

9 The largest telescope in the world, at Mount Palomar, has photographed galaxies estimated to be about **5,000,000,000** light years away.

10 The distance travelled by light in one years is about **10 million million kilometres.**

The Mount Palomar telescope is housed in this building.

Longitude and Time

In 1894 Marconi transmitted the first words sent by radio—from one room to another. In the second half of the twentieth century sounds and pictures are relayed across the world every day. Such events as a royal visit to Australia, World Cup football in Mexico and astronauts landing on the moon are shown on our television screens as they happen.

We know from these programmes that in most other parts of the world the time is different from our own. Why should this be so?

You will remember that the earth is spinning round on its axis from west to east, so that places to the east have their sunrise before us, and places to the west have their sunrise after us. The length of our day is the time it takes the earth to spin round once. This means that:

> In 24 hours the earth turns through 360°
> therefore in 1 hour it turns through 15°

L = London
C = Calcutta
Ch = Chicago

If you look at the picture of a globe you will see a number of circles running from north to south, passing through the poles. These are called **meridians of longitude**.

The longitude of Greenwich, London, is 0°; Calcutta 90° to the east of Greenwich is 6 hours in front of Greenwich Mean Time; Chicago, about 90° to the west of Greenwich, is 6 hours behind Greenwich Mean Time.

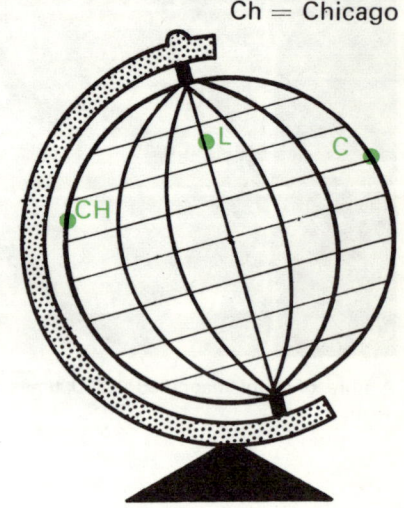

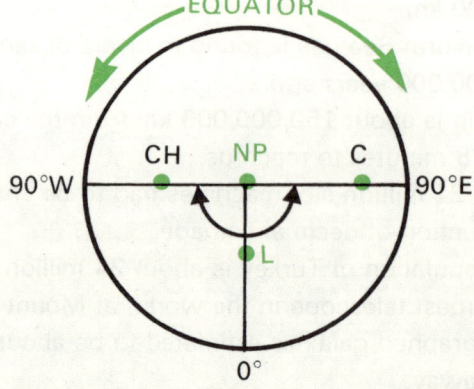

Now let us view the globe as an astronaut might view the earth from above the North Pole. The North Pole would be the centre of the circle, the Equator would be the circumference and each meridian would look like a radius. We can now see how the angles of longitude are fixed, either east or west of Greenwich.

Compare the positions of London, Calcutta and Chicago on the two pictures.

Exercises

A   Copy and complete the following sentences:

1 The earth is revolving on its axis from —— to ——.

2 Places to the —— of us are in front of our time, places to the —— of us are behind us in time.

3 In 60 minutes the earth revolves through —— degrees.

4 In 4 minutes the earth revolves through —— degree.

5 G. M. T. stands for —— —— ——.

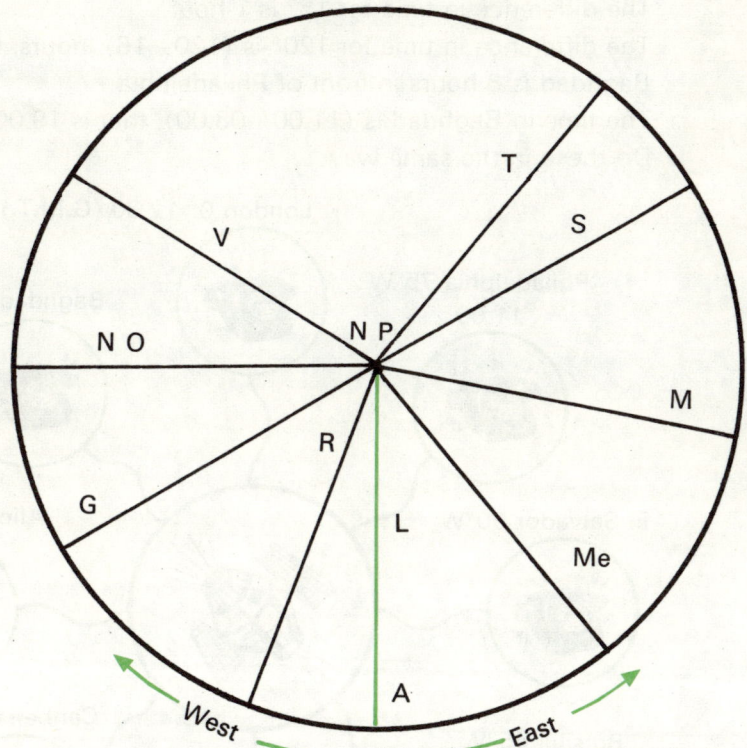

Key	
N P	North Pole
R	Reykjavik
N O	New Orleans
V	Vancouver
G	Georgetown
Me	Mecca
M	Madras
S	Shanghai
T	Tokyo
A	Accra
L	London

B   The diagram shows a view of the northern hemisphere as it might be seen from directly above the North Pole. Use a protractor to find the longitude of each of the cities, and then use your answers to find the cities on a map of the world. Explain why the time is the same in Accra as it is in London.

C   If it is noon G.M.T., what will be the time at the following places? The first example has been done for you.

Example:

Leningrad is 30° east of Greenwich.

The difference in time for 15° is 1 hour.

The difference in time for 30° is (30÷15) hours, that is 2 hours.

Leningrad is 2 hours in front of Greenwich.

The time in Leningrad will be (12.00+02.00), that is 14.00.

1 Leningrad 30° E

2 Kinshasa 15° E

3 Los Angeles 120° W

4 Manila 120° E

5 Montreal 75° W

6 New Britain 150° E

7 Sverdslovsk 60° E

8 Phnom Penh 105° E

9 Zaragoza 0°

10 San Salvador 90° W

43

D   The picture shows a number of cities linked by radio-telephone via a communications satellite.

For some of the cities the time is given when the person is using the telephone. You have to work out the time in the city at the other end of the line.

Example:

Philadelphia talking to **Baghdad**

The difference in longitude is $75° + 45° = 120°$.

The difference in time for $15°$ is 1 hour.

The difference in time for $120°$ is $(120 \div 15)$, hours, that is 8 hours.

Baghdad is 8 hours in front of Philadelphia.

The time in Baghdad is $(11.00 + 08.00)$, that is 19.00.

Do these in the same way:

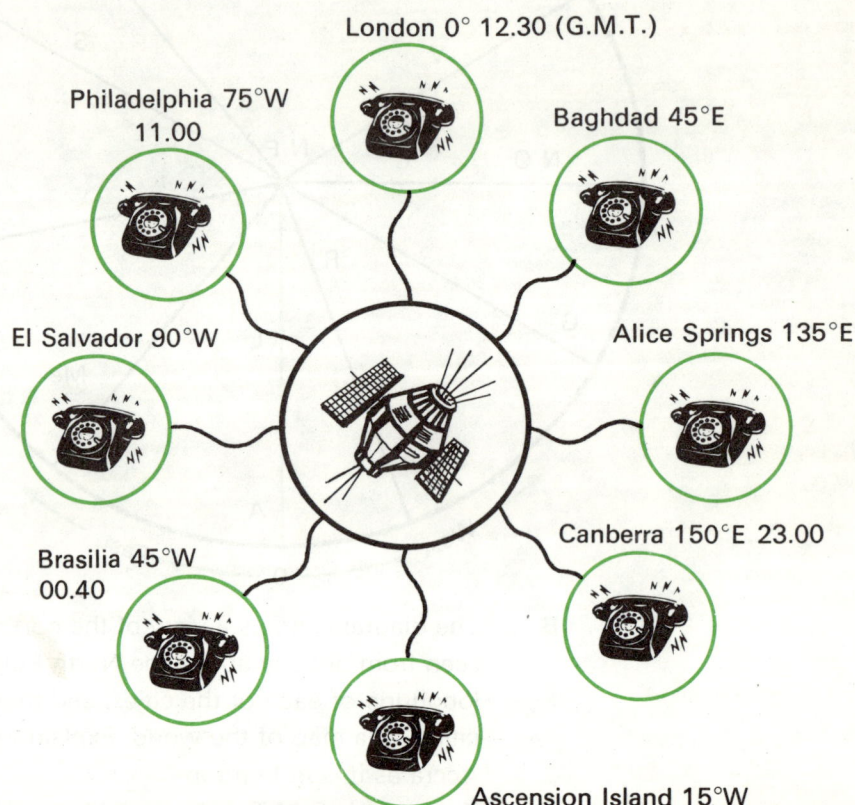

1 London talking to Ascension Island
2 Canberra talking to Baghdad
3 Brasilia talking to Alice Springs
4 London talking to El Salvador
5 Philadelphia talking to Ascension Island
6 Brasilia talking to Baghdad
7 Canberra talking to Ascension Island
8 London talking to Alice Springs

Things to do

Some large countries are divided into 'time zones'. Find out about the time zones in the U.S.A.

Find out what is meant by the International Date Line.

Making Sure—Three

A 1 List the members of the following sets:

    (a) {odd numbers between 38 and 58}

    (b) {prime numbers between 40 and 80}

    (c) {multiples of 9 with two digits}

    (d) {100, 1,000, 10,000, 100,000 written as powers of 10}

    (e) {numbers between 2 and 100 which are divisible both by 3 and by 11}

    (f) {multiples of 25 with three digits and each digit less than 6}

  2 If A = {3, 6, 9, 12}   C = {9, 18, 27, 36}

      B = {6, 12, 18, 24, 30}   D = {12, 24, 36}

    list the members of the following union sets:

    (a) A ∪ B   (c) B ∪ D   (e) C ∪ D

    (b) A ∪ C   (d) B ∪ C   (f) A ∪ B ∪ C

  3 List the members of the individual sets and the union sets represented by the following diagrams:

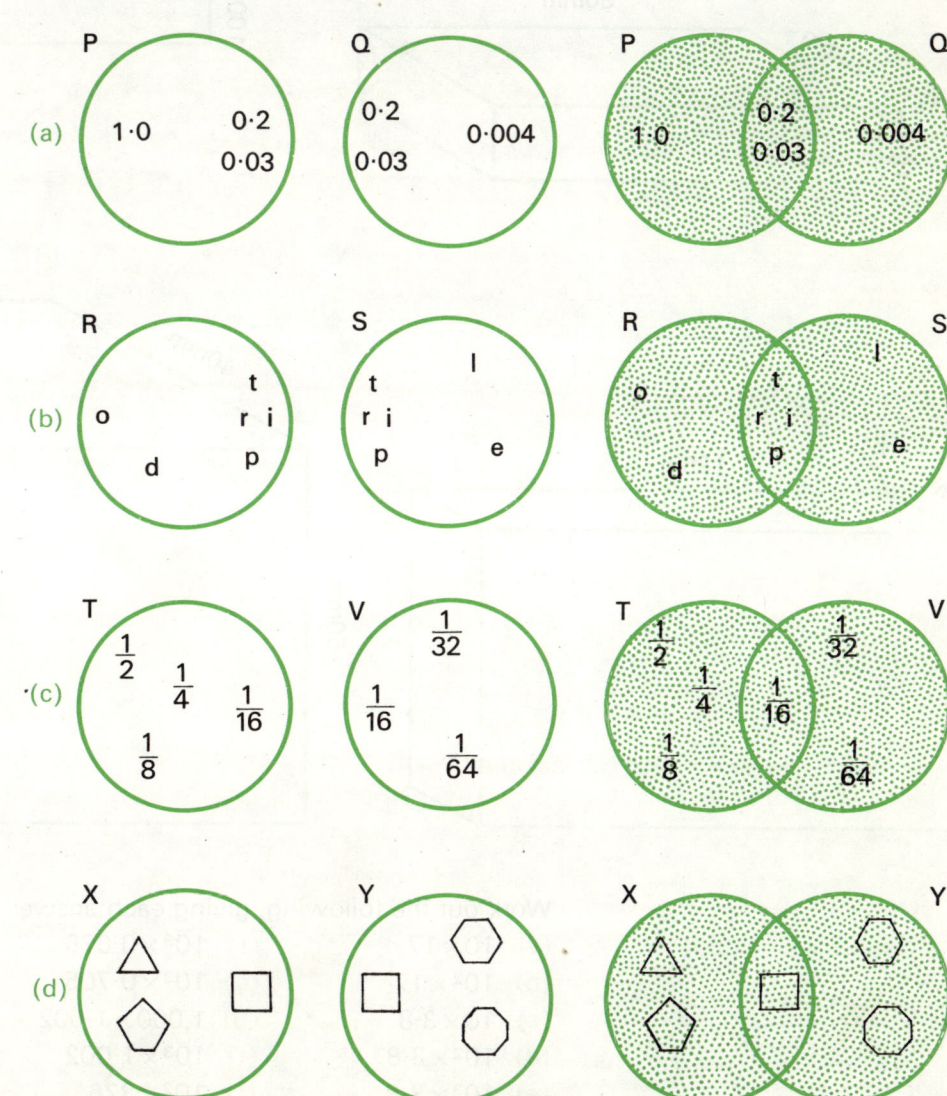

45

B 1 Write down these numbers and then:

   (i) underline the noughts which alter the place and value of any of the other digits,

  (ii) cross out the noughts which do not make any difference to the number.

(a) 407	(f) 03·72	(k) 7,040·600
(b) 470	(g) 400·009	(l) 24,000·002
(c) 047	(h) 004·900	(m) 007·007
(d) 5·09	(i) 94·030	(n) 9,000·9000
(e) 6·070	(j) 3,900·80	(o) 705,006·090

2 Find the volumes of these rectangular prisms and give each answer as a power of 10.

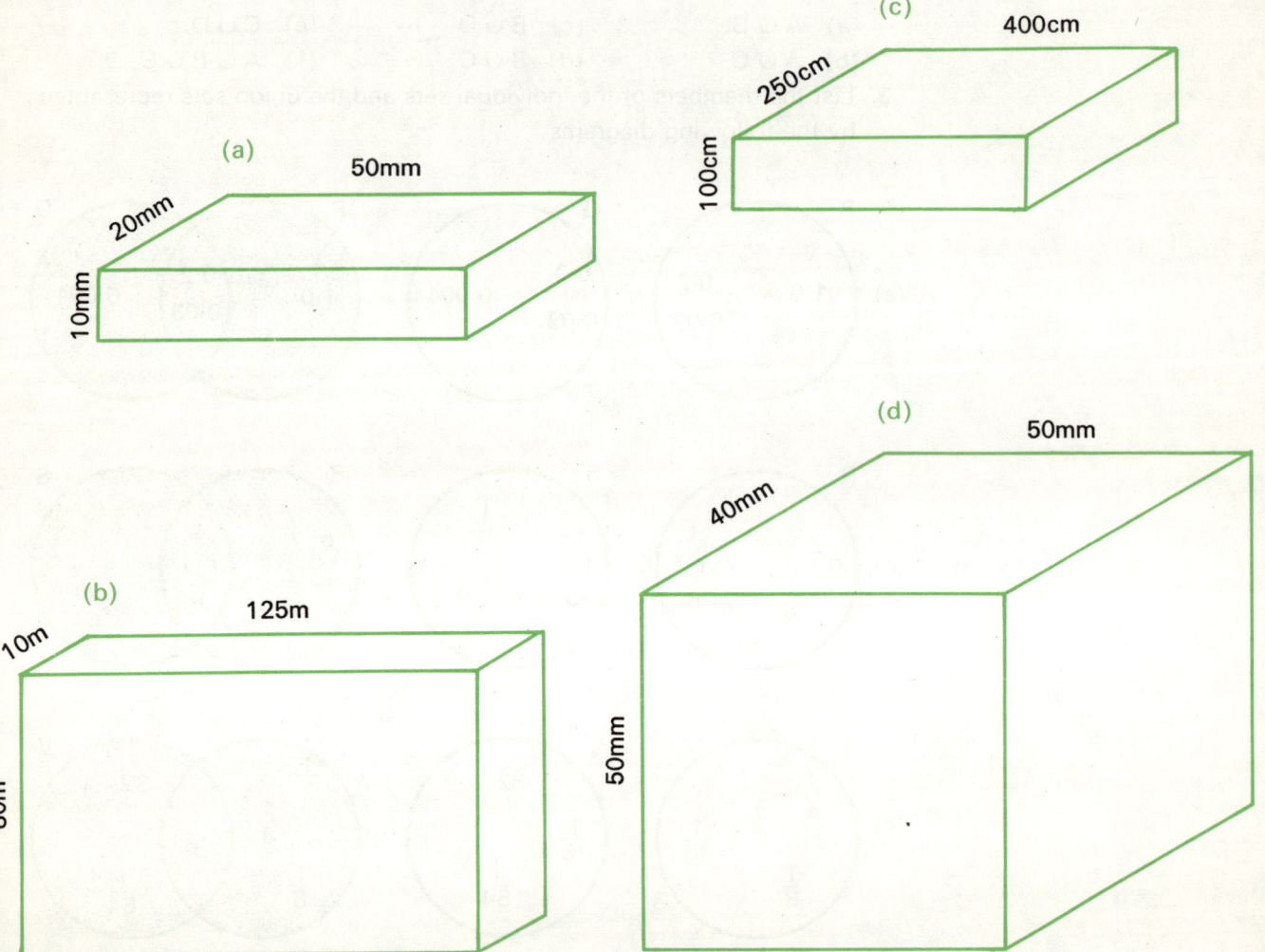

(a) 50mm, 20mm, 10mm

(b) 125m, 10m, 80m

(c) 400cm, 250cm, 100cm

(d) 50mm, 40mm, 50mm

3 Work out the following, giving each answer as an ordinary number:

(a) 10×17	(f) $10^3 \times 0.075$	(k) $10^4 \times 0.003$
(b) $10^2 \times 17$	(g) $10^3 \times 0.705$	(l) $10^2 \times 30.09$
(c) 10×3.8	(h) $1,000 \times 1.002$	(m) $10^5 \times 124$
(d) $10^2 \times 3.8$	(i) $10^3 \times 1.002$	(n) 10×10^2
(e) $10^3 \times 3.8$	(j) $10^2 \times 326$	(o) $10^2 \times 10^4$

4 Express each of the following numbers in standard form.
  (a) A bullet leaves the barrel of a rifle at a speed of about **1,300 km** an hour.
  (b) There are over **20,000** species of bees.
  (c) Mars is about **54,000,000 km** from the Earth.
  (d) Dinosaurs' eggs have been found which date from about **75,000,000 years** ago.
  (e) In 1967 Britain's exports were worth about **£5,000,000,000**.
  (f) At present there are about **3,600,000,000** people in the world.

C 1 If it is three o'clock in the afternoon G.M.T. what will be the time in the following places? Give your answers in the 24 hour clock system.

  (a) Montreal 75° W   (e) Mogadishu 45° E
  (b) Tripoli 15° E   (f) Cairo 30° E
  (c) New Orleans 90° W   (g) Singapore 105° E
  (d) Kangaroo Island 135° E (h) Rio de Janeiro 45° W

Find the above places in an atlas and say what countries they belong to.

2 (a) Make an accurate copy of the diagram below.
  (b) Each letter is the first letter of the name of a city. Use a protractor and an atlas to find out what the cities are.

NP = North Pole   L = London

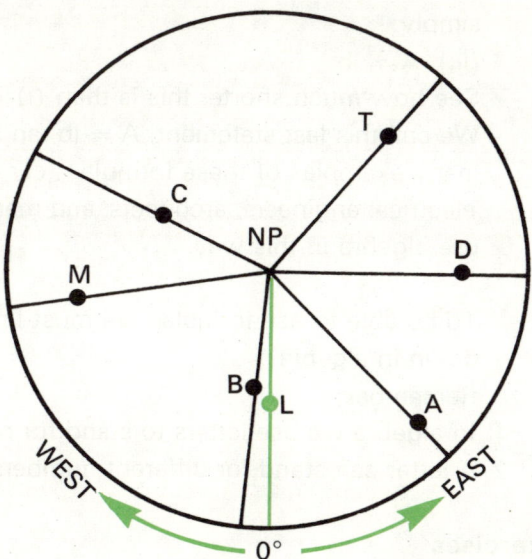

3 Make up a similar diagram to the one above, marking in six different cities. See if your neighbour can discover which cities you have chosen.

12 Algebra

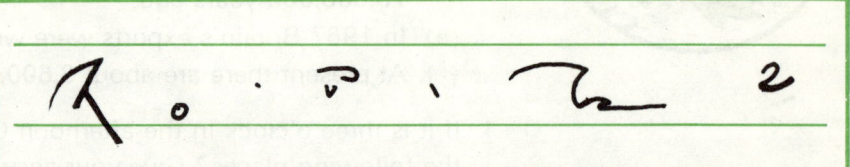

Have you any idea what the above writing means? Probably not, but you may have guessed that it is shorthand. The sentence reads:

'Algebra is a kind of mathematical shorthand.'

Here is a statement about the area of a rectangle:

(i) To find the area we multiply the length by the breadth.

Here is a shorter way of saying the same thing:

(ii) Area = length × breadth.

There is a still shorter way. We can use algebra and say that 'A' is to stand for 'area', 'l' for 'length' and 'b' for 'breadth'. Now we can write 'A = l×b'; but in algebra we can leave out the × sign and write simply:

(iii) A = lb

See how much shorter this is than (i); but it means the same.

We call this last statement, A = lb, an algebraic formula. There are many examples of these formulae: car designers, insurance men, electrical engineers, architects, and many others in business and industry use algebra in this way.

To be able to use formulae we must first learn how to write things down in algebra.

Remember:

1 In algebra we use letters to stand for numbers.

2 A letter can stand for different numbers in different examples.

Exercises

A   Here are some examples of how statements can be shortened:

$4 \times 5 \longrightarrow 20$

$4 \times a \longrightarrow 4a$

$a \times b \longrightarrow ab$

Now write the following numbers in a shorter form.

1 2×5	6 $7 \times c$	11 $2 \times 3 \times c$	16 $2 \times 7 \times e \times f$
2 $2 \times a$	7 $5 \times d$	12 $a \times b \times c$	17 $1 \times e \times f$
3 1×6	8 6×8	13 $3 \times a \times b$	18 $2 \times 3 \times p \times q$
4 $1 \times a$	9 $6 \times y$	14 $5 \times c \times d$	19 $2 \times p \times 3 \times q$
5 $a \times c$	10 $c \times d$	15 $c \times d \times e$	20 $5 \times a \times b \times c \times d$

B Find the perimeters of the following figures, expressing your answers in the shortest possible form.

Example:

3a

3a   3a

3a

$$P = 4 \times 3a$$
$$= 12a$$

1 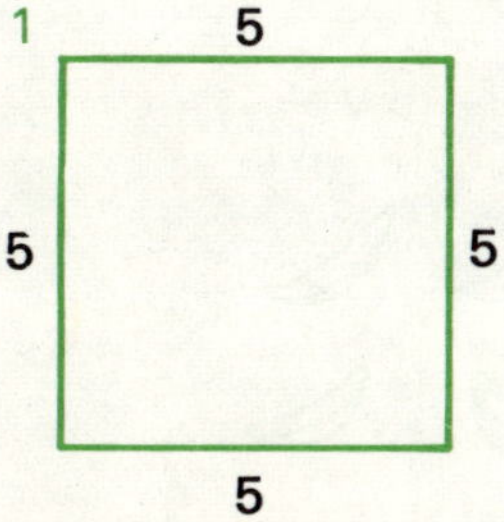
5
5   5
5

2 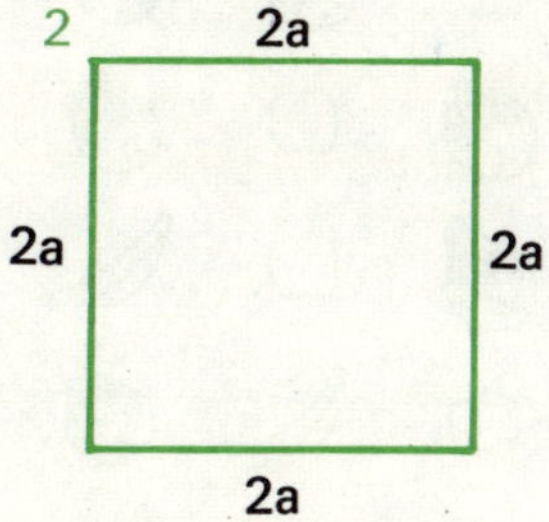
2a
2a   2a
2a

3 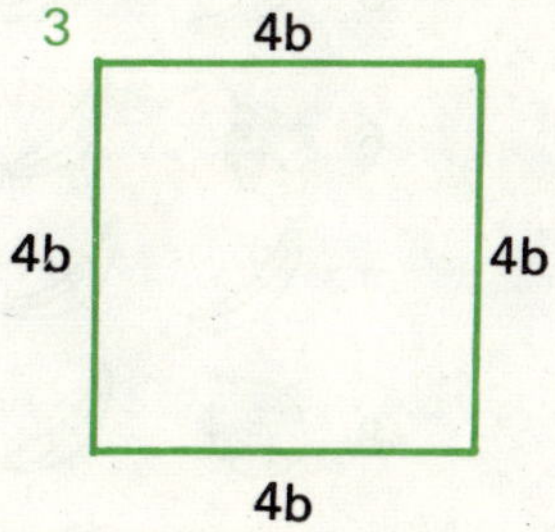
4b
4b   4b
4b

4
$\frac{1}{2}c$
$\frac{1}{2}c$ $\frac{1}{2}c$
$\frac{1}{2}c$

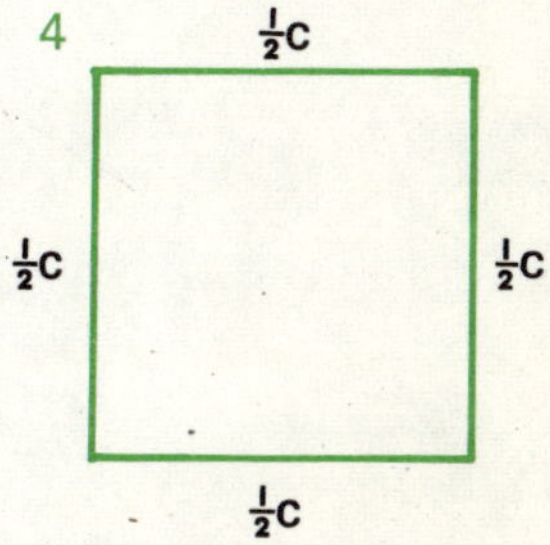

5
6m   6m
6m

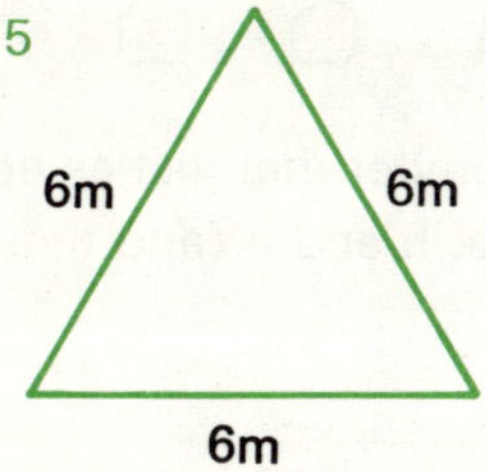

6
10y   10y
10y

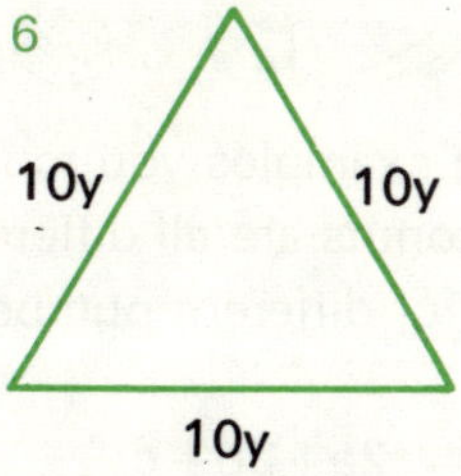

7
9d   9d
9d

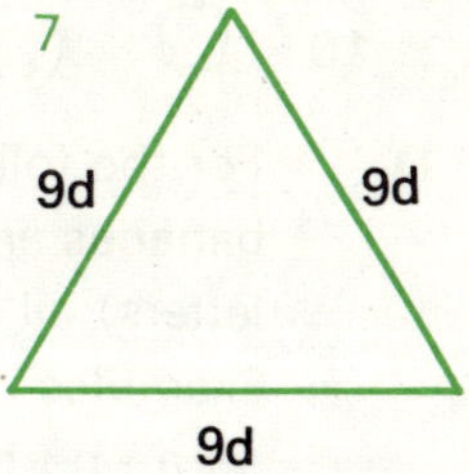

8
$\frac{1}{3}a$ $\frac{1}{3}a$
$\frac{1}{3}a$

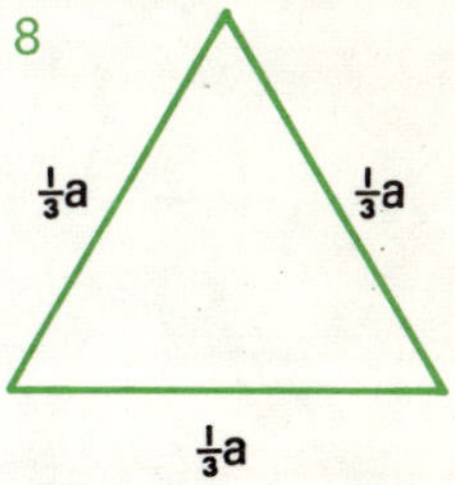

9
ab
ab   ab
ab

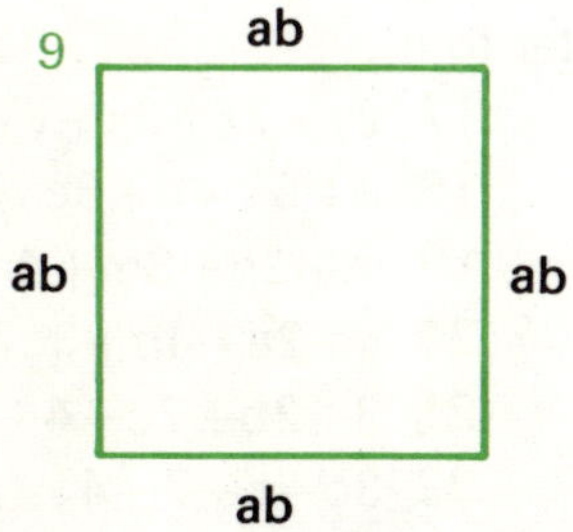

10
2xy   2xy
2xy

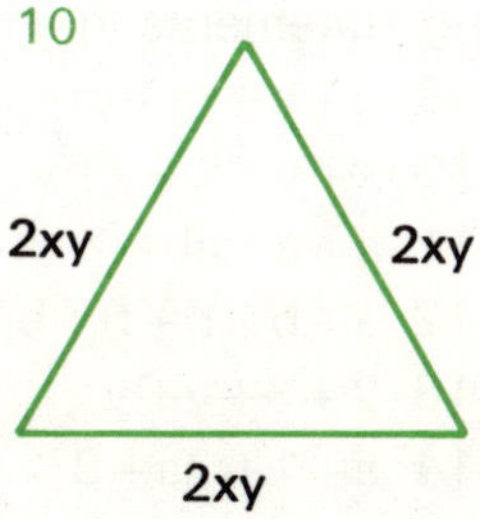

11
4c
4c   4c
4c   4c
4c

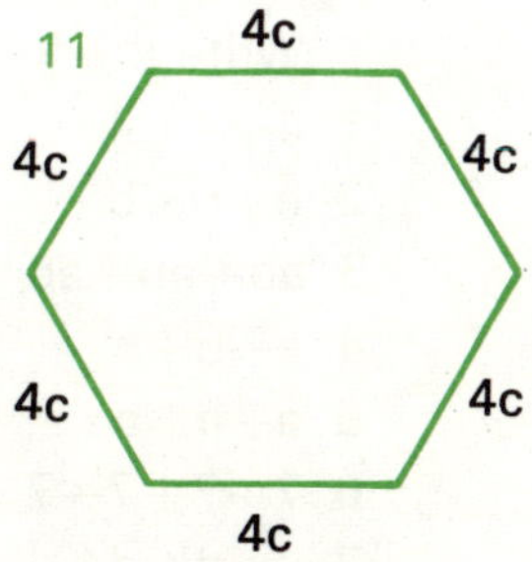

12
3ab   3ab
3ab   3ab
3ab

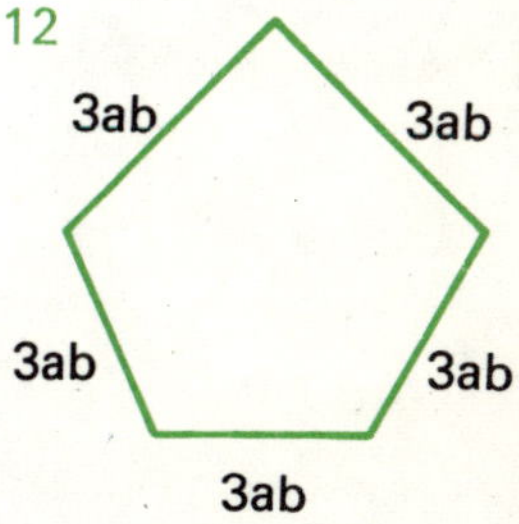

C In this exercise let **a** stand for an apple,
b stand for a banana,
and **c** stand for a coconut.
Now study these two examples, which show another way in which statements can be shortened.

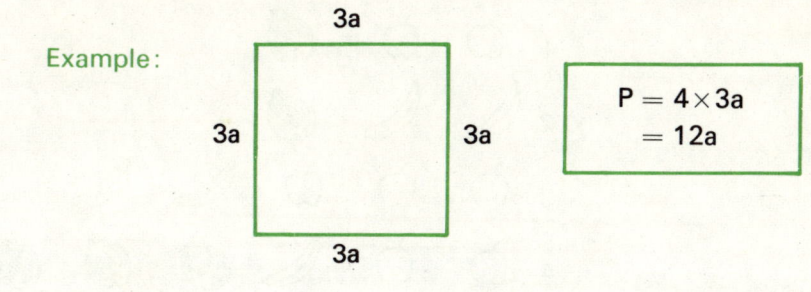

◯ + ◯ + 〰 ⟶ a+a+b → 2a+b

◯ + 〰 + 〰 + 🥥 + 🥥 + 🥥 → a+b+b+c+c+c → a+2b+3c

49

You should now be able to write an algebraic statement to represent each of the following. Remember to express your answers in the shortest possible form.

D For the following examples you must remember that just as **apples**, **bananas** and **coconuts** are all different so **a**, **b** and **c** (and the other letters) all stand for different numbers.

Examples:

a+a+b+b+c → 2a+2b+c

x+x+x+y+y+z → 3x+2y+z

Write the following statements in a shorter form:

1 4+4+4
2 b+b+b
3 ab+ab+ab
4 a+b+b
5 a+a+b
6 7+7+7+7
7 c+d+d+d
8 5+a+a

9 n+n+m+m+m
10 x+y+x+y+y
11 a+a+ab
12 a+b+b+b+b+b
13 2+3+y+y
14 a+1+2a+3
15 a+2a+b+3b
16 a+a+b+c

17 2y+2x+3x+y
18 6+5+4b+3c
19 m+2n+3m+4n
20 a+2a+4b+3c+2c
21 3+2b+2c+4+c
22 3c−c+7−4
23 a+2a+2b−b
24 ab+ab+bc+bc+bc

E Find the perimeters of the following figures, expressing your answers in the shortest possible form. Give the name of each figure if you can.

Example:

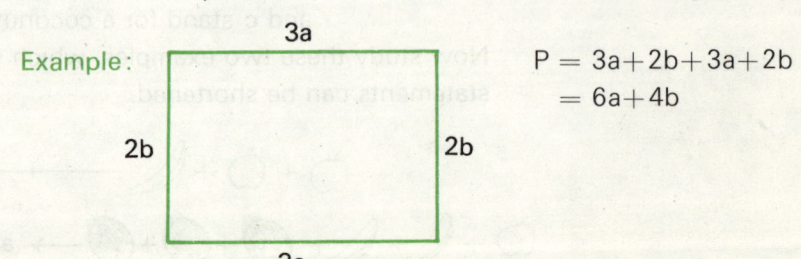

P = 3a+2b+3a+2b
  = 6a+4b

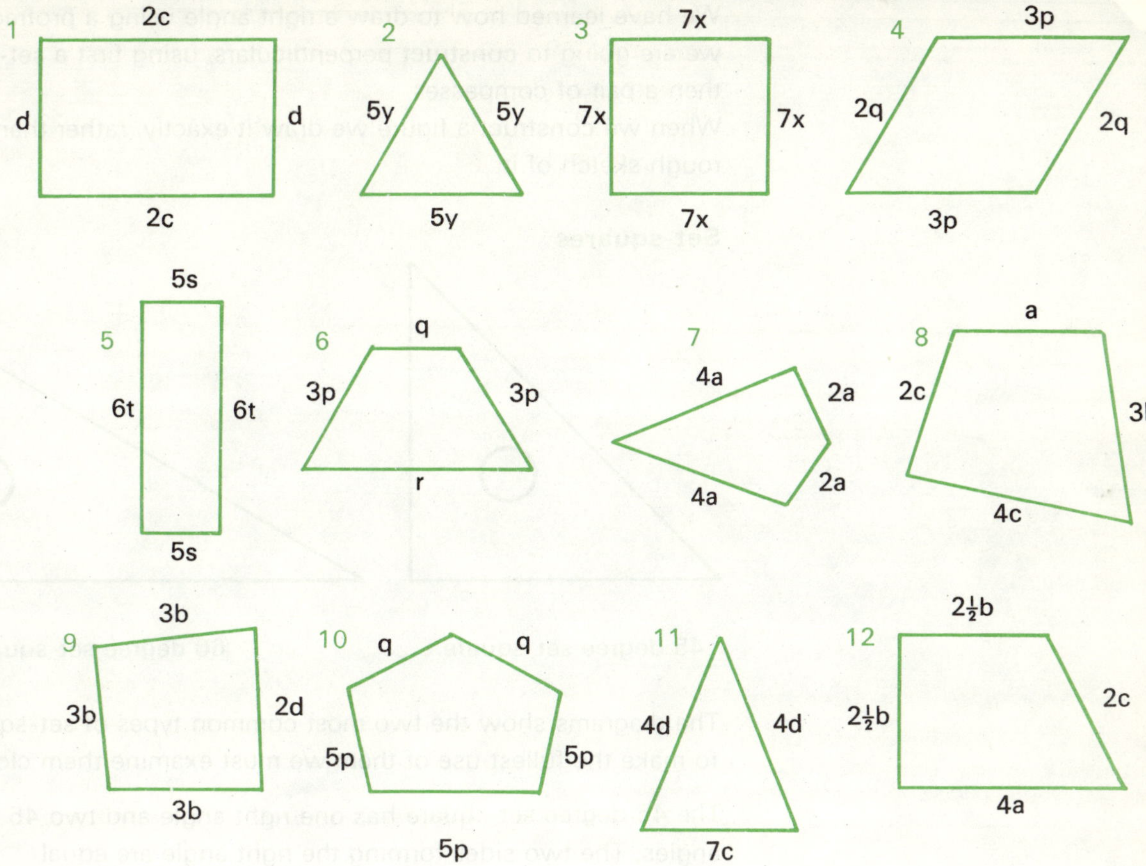

1

2c / d / d / 2c

2

5y / 5y / 5y

3

7x / 7x / 7x / 7x

4

3p / 2q / 2q / 3p

5

5s / 6t / 6t / 5s

6

q / 3p / 3p / r

7

4a / 2a / 4a / 2a

8

a / 2c / 2c / 3b / 4c

9

3b / 3b / 2d / 3b

10

q / q / 5p / 5p / 5p

11

4d / 4d / 7c

12

$2\frac{1}{2}b$ / $2\frac{1}{2}b$ / 2c / 4a

Perpendiculars

Why is the tower in the picture famous? If you did not know about it beforehand, you will probably have already guessed that it is because the walls are not upright. Indeed it is known as the 'Leaning Tower of Pisa'.

Normally, if the ground is level, or horizontal, the walls of a building make a right angle with the ground, as shown in the drawing on the right.

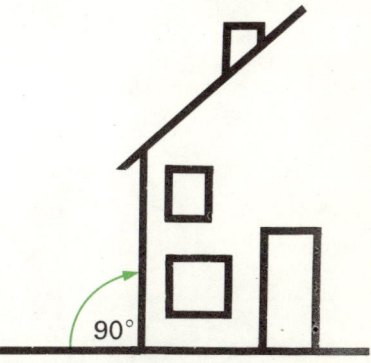

90°

> When two straight lines meet at right angles they are said to be **perpendicular** to each other.

Here are some more examples of perpendicular lines.

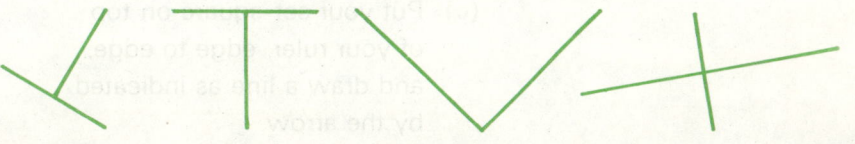

We have learned how to draw a right angle using a protractor. Now we are going to **construct** perpendiculars, using first a set-square and then a pair of compasses.

When we **construct** a figure we draw it exactly, rather than just make a rough sketch of it.

Set-squares

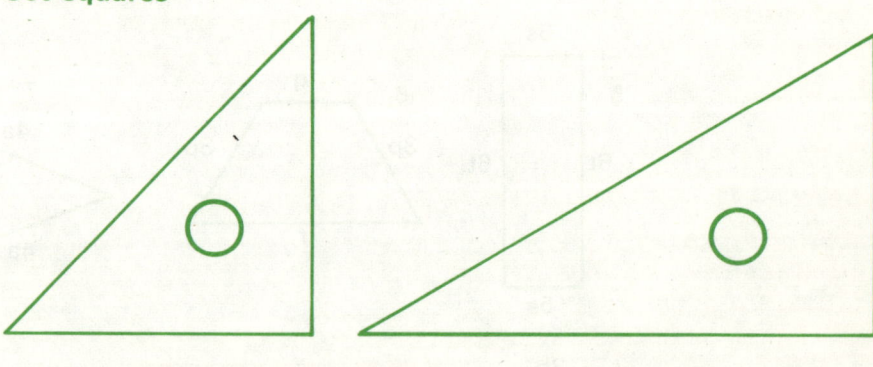

45 degree set-square   60 degree set-square

The diagrams show the two most common types of set-square, and to make the fullest use of them we must examine them closely.

The 45 degree set-square has one right angle and two 45 degree angles. The two sides forming the right angle are equal.

The 60 degree set-square has one right angle, one 60 degree angle and one 30 degree angle. The longest side is twice as long as the shortest side.

To construct a line CD perpendicular to a line AB:

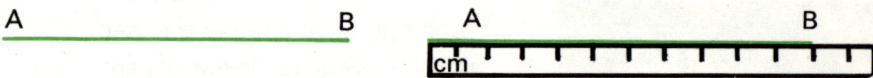

(a) Draw a line AB any length.

(b) Place the edge of your ruler along AB.

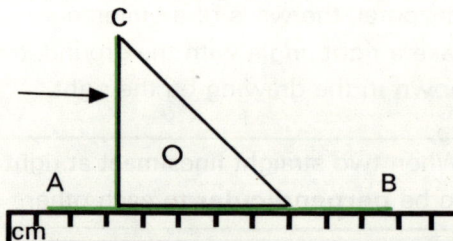

(c) Put your set-square on top of your ruler, edge to edge, and draw a line as indicated by the arrow.

CD is now perpendicular to AB.

Compasses

To draw a line LM perpendicular to a line RS:

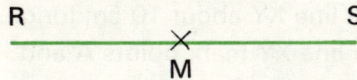

(a) Draw a line RS any length, and mark on it a point M.

(b) With centre M, and any convenient radius, draw arcs to cut RS at P and Q.

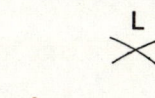

(c) With centres P and Q, draw arcs to cut at L. The radius must be greater than the distance PM.

(d) Draw a line from M to pass through the point L.

Exercises

A 1 (a) Find the total of the three angles of a 45 degree set-square.

(b) Find the total of the three angles of a 60 degree set-square.

2 Use a 45 degree, and a 60 degree, set-square to draw the following angles:

Example: 105°

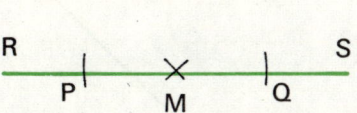

(a) 75° (b) 135° (c) 150° (d) 120° (e) 15° (f) 165°

3 Use the method described either,

(i) to draw the plan of a 45 degree set-square in your book, or

(ii) to make your own set-square in cardboard.

(a) Draw a line XY about 10 cm long.

(b) On the line XY mark points A and B 6 cm apart.

(c) Use compasses to erect a perpendicular at B, and draw BC 6 cm long.

(d) Join A to C.

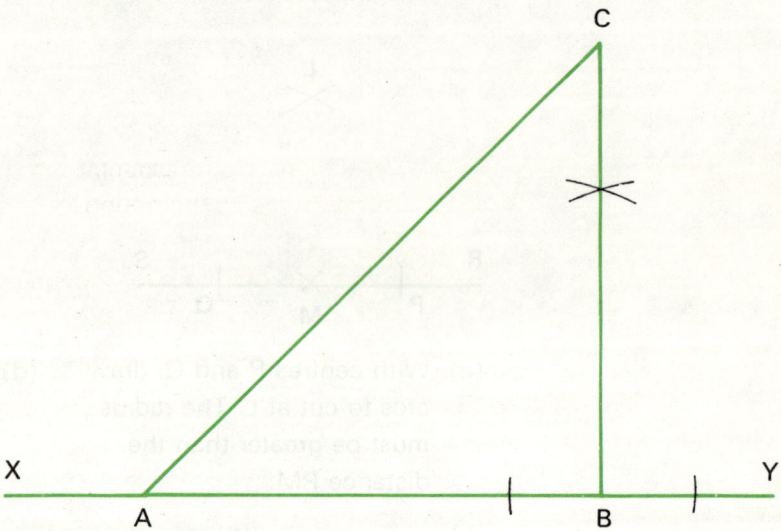

4 Use the method described either,

(i) to draw the plan of a 60 degree set-square in your exercise book, **or**

(ii) to make your own 60 degree set-square in cardboard.

(a) Draw a line XY about 12 cm long.

(b) On the line XY mark a point B, about 2 cm from Y.

(c) Erect a perpendicular at B, and draw BC 5 cm long.

(d) With centre C, and radius 10 cm, draw an arc to cut the line XY at A.

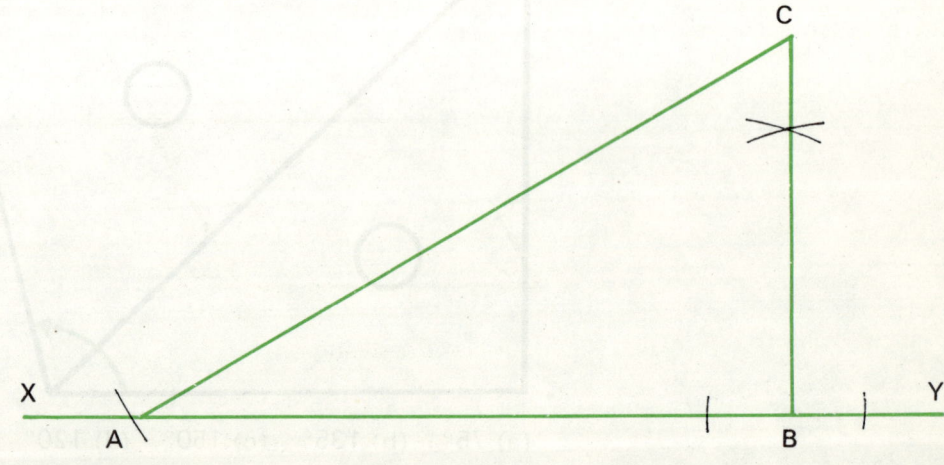

5 Make a right angle by folding a piece of paper as shown, and then use it to test the right angles on your drawings or set-square.

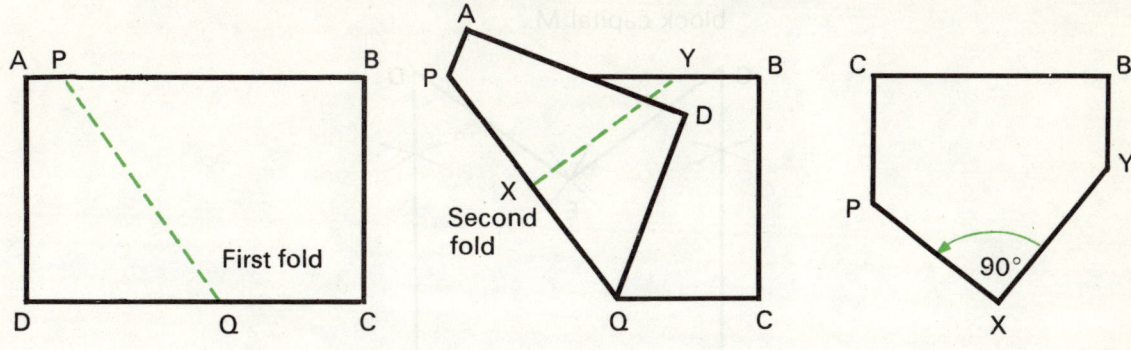

First fold
Second fold
90°

B Use a **set-square** for the examples in this section.
1 (a) Draw a line AB 10 cm long.
(b) Mark a point D, on AB, 4 cm from A.
(c) Draw a line CD 5 cm long perpendicular to AB.

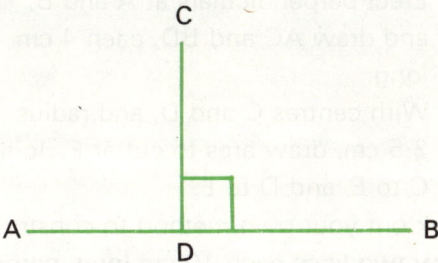

2 Construct these five block capitals. Notice that some lines are parallel, and some lines are perpendicular to each other. Make your letters the same size as these.

3 Construct the two figures shown below.

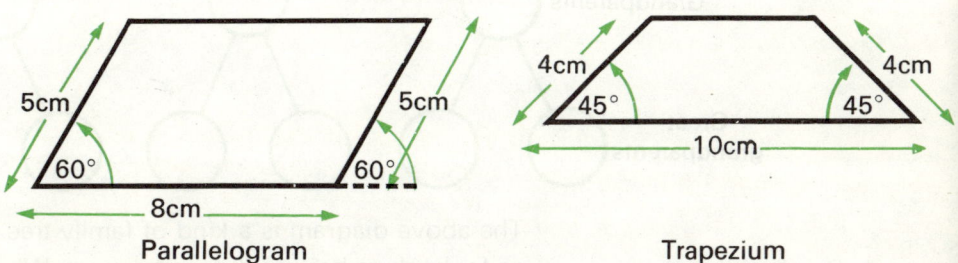

5cm   5cm
60°   60°
8cm
Parallelogram

4cm   4cm
45°   45°
10cm
Trapezium

Find out the definitions of a parallelogram and a trapezium, and write them in your exercise book.

C Use **compasses** for the examples in this section.

1 Use the method outlined to construct a block capital M.

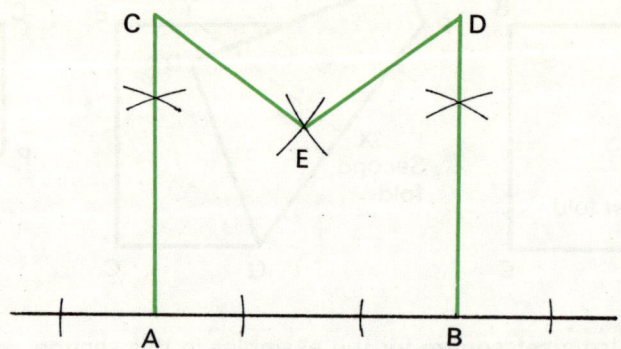

(a) Draw a line about 8 cm long, and mark on it two points, A and B, 4 cm apart.

(b) Erect perpendiculars at A and B, and draw AC and BD, each 4 cm long.

(c) With centres C and D, and radius 2·5 cm, draw arcs to cut at E. Join C to E, and D to E.

2 Work out your own method to construct a capital A.

3 Draw two lines each 10 cm long, perpendicular to each other and bisecting each other.
Use these two lines as the basis of a pattern, and then colour it in.

Computer Numbers

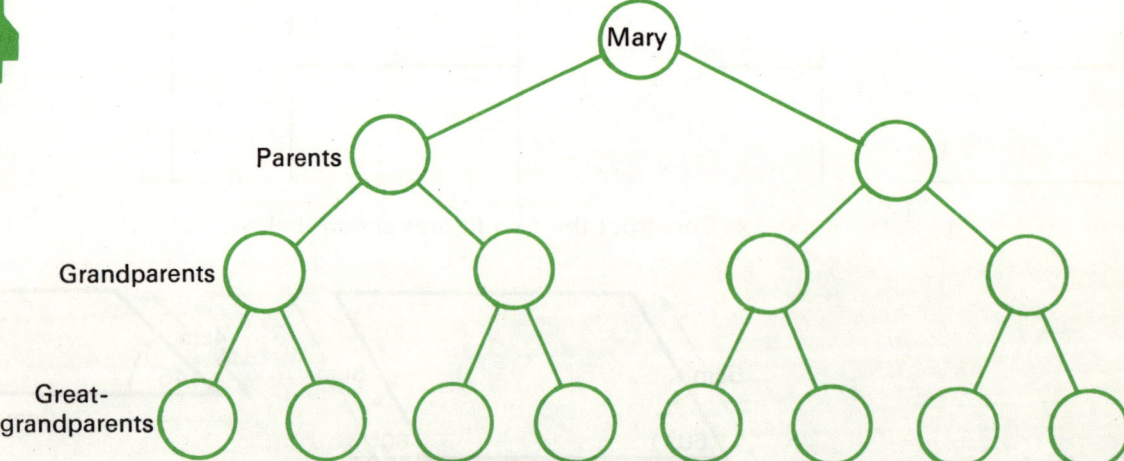

The above diagram is a kind of family tree. It shows Mary's relations as far back as her great-grandparents. What numbers should go with each of the following:

Mary, **Parents,** **Grandparents,** **Great-grandparents?**

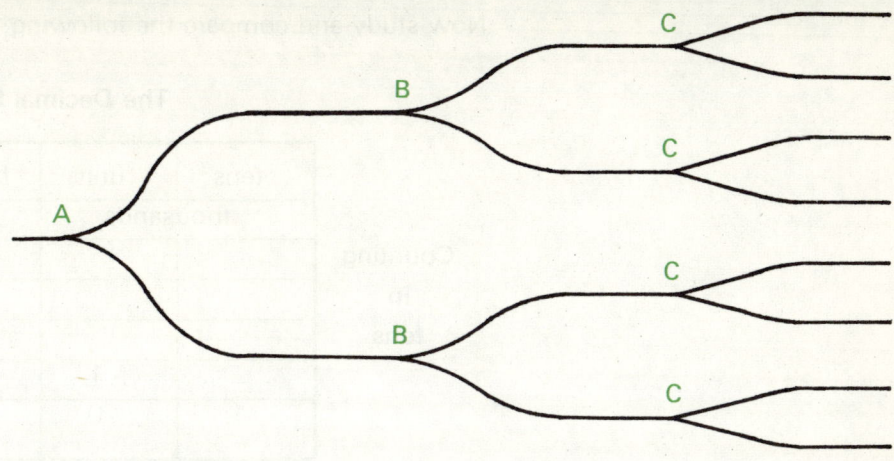

Now look at the diagram showing the branch lines of a railway marshalling yard. The points have been labelled A, B and C.

How many lines are there:

at the start,   after points A,   after points B,   after points C?
Did you get these numbers: **1, 2, 4** and **8** in each case?
What would be the next two numbers if each of the diagrams were continued?
There are many examples of these 'two way' systems, but by far the most important is the modern computer. Computers are transforming our lives; such varied activities as designing aeroplanes, tracking down criminals, forecasting the weather, controlling traffic and guiding spacecraft are just a few examples of the work they do.

Part of the control centre during a space flight.

Computers can work out calculations much more quickly than we can; but in one way, at least, we are more clever: we can count in **tens**, and they have to count in **twos**. This is because a computer works through a large number of transistors, which can be switched either 'on' or 'off'. Only **two** things can happen (switch 'on' or switch 'off'), and so computers must use a **base 2** system of counting, which we usually call the **Binary System**.
Let us compare the binary system with our decimal system of numbers.

Decimal	Binary
Counts in **tens**	Counts in **twos**
Uses **ten** digits:	Uses **two** digits:
0, 1, 2, 3, 4,	0 and 1
5, 6, 7, 8, 9	
1 stands for a whole **one**	1 stands for a whole **one**
10 stands for a group of **ten**	10 stands for a group of **two**
100 stands for **ten** groups of **ten** (a hundred)	100 stands for **two** groups of **two** (four)

Now study and compare the following tables.

The Decimal System

tens	units	hundreds	tens	units
thousands		units		
				1
			1	0
		1	0	0
	1	0	0	0
1	0	0	0	0

Counting in tens

The Binary System

sixteens	eights	fours	twos	units
				1
			1	0
		1	0	0
	1	0	0	0
1	0	0	0	0

Counting in twos

Here are the first sixteen counting numbers in the binary system.
The key numbers from the above table have been underlined.

one	1	five	101	nine	1001	thirteen	1101
two	10	six	110	ten	1010	fourteen	1110
three	11	seven	111	eleven	1011	fifteen	1111
four	100	eight	1000	twelve	1100	sixteen	10000

Exercises

A Explain the meaning of these words and then say how they are connected. Use a dictionary if necessary.

1 bicycle   4 bisect   7 bilingual
2 biplane   5 biannual   8 biennial
3 biped   6 binoculars

B 1 Copy the first sixteen counting numbers in the binary system and then continue as far as thirty-two.

2 Say how we can tell whether a binary number is odd or even.

3 If you put a nought at the right hand side of a decimal whole number what number have you multiplied by?

4 If you put a nought at the right hand side of a binary number what number have you multiplied by?

C Look at each of the following diagrams and say how many different routes there are from the top pin to any one of the bottom pins, following the direction of the arrows. This means you cannot go sideways. Give your answer in the binary system.

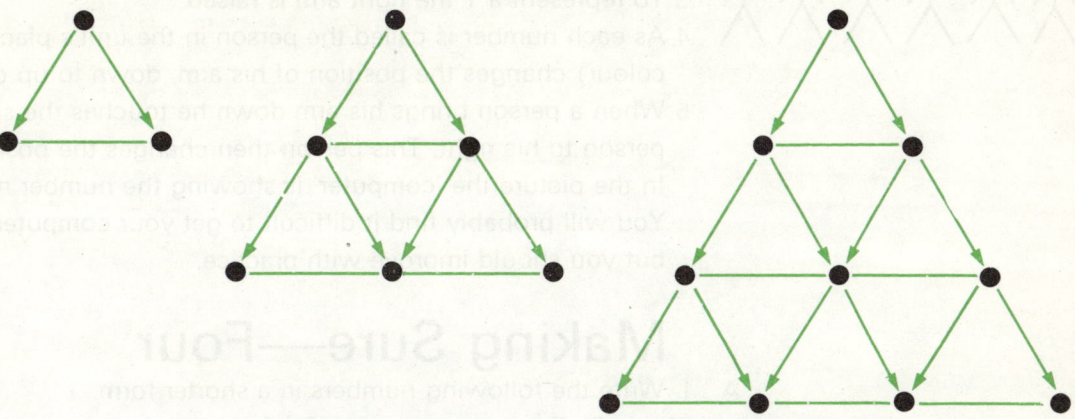

D Draw a figure similar to the ones above, this time with 5 rows of pins. Once again work out the number of routes from top to bottom. How many routes would there be with 6 rows and 7 rows of pins?

E Give in words the value of each of the figures underlined in these binary numbers.

1 10_1_	4 _1_111	7 _1_00001	10 111_1_1
2 1_0_0	5 1_0_001	8 101_0_11	11 1_1_0011
3 101_0_	6 10_1_01	9 110_1_11	12 1_1_001100

F When we are adding numbers in the binary system we must remember that $1+1 = 10$, so we put 0 down and 'carry' 1.

Examples:

```
    one  →     1        one   →     1        three →    11
 + two  → + 10       +three → +11        +three →  +11
  ─────────────      ──────────────       ─────────────
  three  →    11        four  →   100         six  →   110
                                   11                   11
```

Notice in the last example that $1+1+1 = 11$.

All the following examples are in binary:

1 100 + 11	5 1000 + 111	9 111 +1001	13 11111 + 1000
2 101 + 1	6 1010 + 1010	10 11 +111	14 10001 + 111
3 110 + 10	7 1111 + 100	11 10110 + 1010	15 11011 +10101
4 10 +10	8 111 +101	12 11011 + 101	16 11111 +10001

G Now see if you can make a human computer work. Here are the instructions:

1. Six members of the class stand in line as shown in the picture.
2. To represent a 0 both arms remain down.
3. To represent a 1 the right arm is raised.
4. As each number is called the person in the unit's place (he is in colour) changes the position of his arm, down to up or up to down.
5. When a person brings his arm down he touches the shoulder of the person to his right. This person then changes the position of his arm. In the picture the 'computer' is showing the number nine (001001). You will probably find it difficult to get your computer working at first but you should improve with practice.

Making Sure—Four

A 1 Write the following numbers in a shorter form.

(a) 3×7 (f) $2 \times 3 \times d$ (k) $6 \times e \times f \times 2$
(b) $3 \times a$ (g) $2 \times a \times b$ (l) $a \times b \times c$
(c) 1×5 (h) $3 \times 5 \times d$ (m) $3 \times a \times b \times c$
(d) $1 \times b$ (i) $3 \times d \times 5$ (n) $p \times 4 \times 2 \times q$
(e) $b \times c$ (j) $2 \times c \times 3 \times d$ (o) $a \times 7 \times b \times c \times d$

2 Find the perimeters of the following figures, expressing your answers in as short a form as possible.

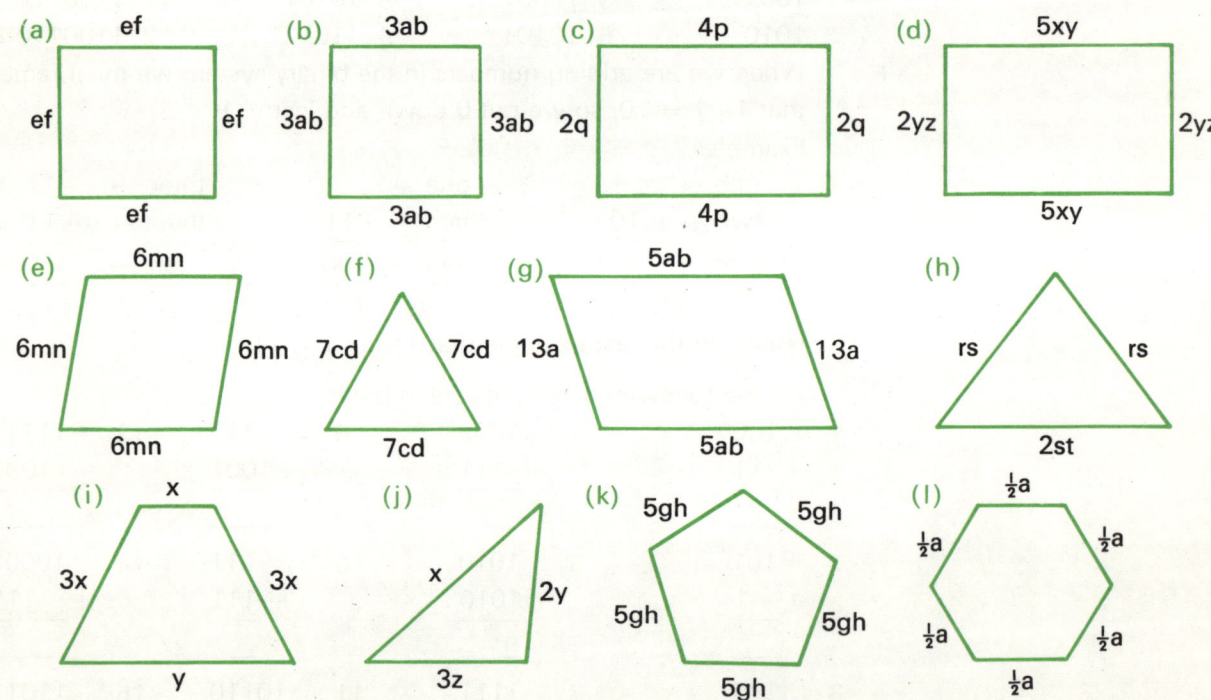

3 Express the following in the shortest possible form.

(a) $3+4+5$ (f) $a+b+2a+3b$ (k) $ab+bc+ab+2ac$
(b) $3a+4a+5a$ (g) $3p+4p+5p$ (l) $2x+3xy+4x+4y$
(c) $bc+2bc+3bc$ (h) $3b+a-2b$ (m) $3a+4b+ab+2ab$
(d) $x+2x+2y+3y$ (i) $4+b-1-b$ (n) $ab+2abc+ab+3abc$
(e) $2+4a+6a$ (j) $2m+3n+7m-n$ (o) $4+3ab-4-abc$

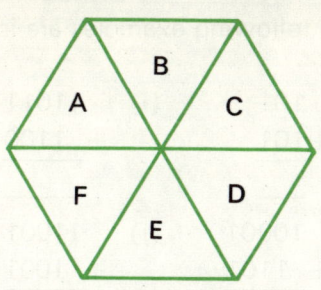
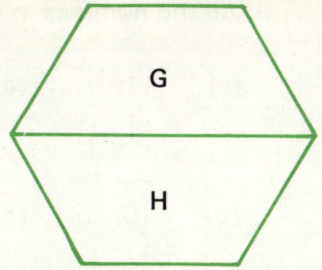
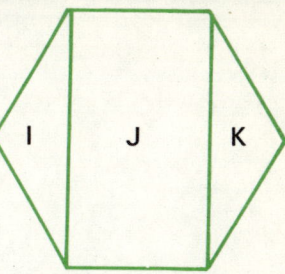

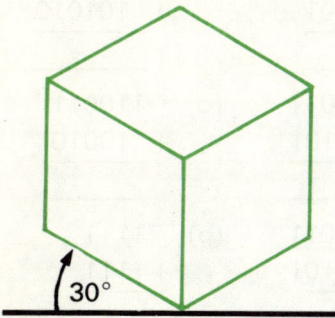

30°

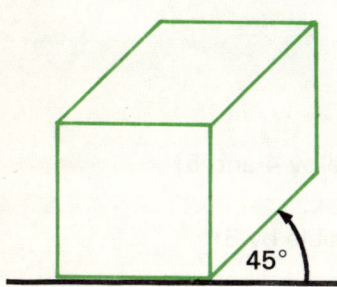

45°

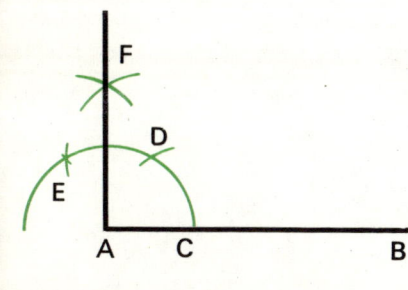

F

D

E

A   C   B

B 1 These drawings show three different ways of dividing up a hexagon. In each case the figures making up the hexagon have been given a letter.

(a) Name the figure for each letter.

(b) Use a 60° set-square to find the sizes of the angles in each figure.

(c) Use a 60° set-square to construct a hexagon with sides measuring 5 cm.

2 Draw these two views of a cube, using a 60° set-square for the first one and a 45° set-square for the second one. Choose your own length for the edges of the cubes.

3 Here is a method for erecting a perpendicular at the end of a line.

Draw a line AB.

With centre A and any suitable radius draw a semicircle to cut AB at C.

With centre C and the **same** radius draw an arc to cut the semicircle at D.

With centre D and the same radius draw an arc to cut the semicircle at E.

Using D and E as centres, and with the same radius, draw arcs to cut at F.

Draw a line from A through F.

(a) Practise drawing perpendiculars using this method.

(b) Use a ruler and a pair of compasses to draw a figure similar to the one shown below. You might make the figure the centre-piece of a picture.

C 1 **(a)** Write out the numbers one to sixty-four in the decimal system.

(b) Write the following 'key' binary numbers alongside the corresponding decimal numbers:

one	**1**	four	**100**	sixteen	**10000**
two	**10**	eight	**1000**	thirty-two	**100000**
		sixty-four	**1000000**		

(c) Write out the remainder of the binary numbers.

2 Give in words the value of each of the figures underlined in these binary numbers.

(a) 1̲0   **(e)** 1̲11011   **(i)** 1̲100011

(b) 101̲   **(f)** 1̲11000   **(j)** 1̲0000

(c) 11̲01   **(g)** 1001̲00   **(k)** 1101̲101

(d) 1̲100   **(h)** 11̲0001   **(l)** 11̲11111

3 All the numbers in the following examples are in the binary system.

(a)	11	(e)	101	(i)	11011	(m)	110001
	+ 1		+101		+ 1100		+110001

(b)	10	(f)	10001	(j)	11001	(n)	1010101
	+11		+ 1101		+ 1001		+ 101010

(c)	101	(g)	1101	(k)	110011	(o)	110011
	+ 10		+ 101		+110101		+ 10010

(d)	110	(h)	110	(l)	111011	(p)	1111
	+ 10		+110		+ 10101		+1111

Revision Exercises

A List the members of the following sets:
1 {multiples of 7 between 6 and 85}
2 {numbers between 19 and 81 exactly divisible by 4 and 5}
3 {square numbers between 0 and 145}
4 {odd numbers between 2 and 34 exactly divisible by 3}
5 {leap years between 1971 and 2001}

B List the intersection sets for the following pairs of sets:
1 A = {1, 2, 3, 5, 7, 11}
   B = {1, 3, 5, 7, 9}
2 C = {5, 10, 15, 20, 25}
   D = {10, 20, 30}
3 E = {d, e, p, o, r, t}
   F = {p, o, s, t, e, r}
4 G = {ab, bc, cd, de}
   H = {ab, cd, ef, gh}
5 I = {0°, 45°, 90°, 135°}
   J = {90°, 135°, 180°, 225°, 270°}

C Use the distributive law to work out the following examples:
1 $(3 \times 5) + (3 \times 5)$ 6 $(8 \times 57) - (8 \times 7)$
2 $(2 \times 2) + (2 \times 18)$ 7 $(6 \times 2) + (6 \times 3) + (6 \times 5)$
3 $(4 \times 27) + (4 \times 23)$ 8 $(10 \times \frac{1}{7}) + (10 \times \frac{2}{7}) + (10 \times \frac{4}{7})$
4 $(4 \times 2\frac{3}{8}) + (4 \times \frac{5}{8})$ 9 $(12 \times 9) + (12 \times 991)$
5 $(11 \times 95) + (11 \times 5)$ 10 $(11 \times 200) + (11 \times 300) + (11 \times 500)$

D Change to metres:
1 7,000 mm   6 11,000 mm
2 700 cm   7 900 cm
3 7 km   8 950 cm
4 3·5 km   9 4·001 km
5 3·05 km   10 12·086 km

E   Change to kilometres:

1. 4,000 m
2. 4,500 m
3. 14,000 m
4. 4,005 m
5. 12,500 m

6. 12,250 m
7. 9,001 m
8. 100,000 m
9. 15,035 m
10. 15,350 m

F

1. $\frac{3}{4}+\frac{4}{7}$
2. $1\frac{3}{7}+2\frac{5}{7}$
3. $\frac{3}{4}+\frac{7}{8}$
4. $4\frac{1}{2}+3\frac{3}{5}$
5. $\frac{1}{4}+\frac{1}{3}+\frac{1}{2}$

6. $2\frac{1}{6}+1\frac{2}{9}$
7. $\frac{1}{5}+\frac{3}{10}+\frac{1}{2}$
8. $1\frac{1}{3}+2\frac{1}{6}+3\frac{2}{9}$
9. $2\frac{1}{4}+1\frac{5}{8}+3\frac{1}{2}$
10. $1\frac{1}{2}+2\frac{2}{5}+4\frac{7}{10}$

G

1. The school team have a goal average of 2·2 for 20 matches. How many goals have they scored altogether?
2. Alan set off on a bicycle ride at 11.50 and arrived home at 15.50. He covered a total distance of 64 km. What was his average speed?
3. A tin of peas weighs 250 g. What is the approximate weight of a carton holding 80 tins? Give your answer in kilograms.
4. The distance round an athletics track is 400 m. How many laps will be needed to complete a 1,500 m race? Your answer should be a mixed number.
5. Barbara saved up at the rate of 15p a week from the first of November to the nineteenth of December. How much did she save altogether?

H   Copy these statements and put the correct number in the box to make the statements true:

1. $4\times7=\square$
2. $4\times0·7=\square$
3. $5\times12=\square$
4. $5\times1·2=\square$
5. $8\times8=\square$

6. $8\times0·8=\square$
7. $8\times0·08=\square$
8. $2\times25=\square$
9. $2\times2·5=\square$
10. $2\times0·25=\square$

I   List the union sets for the following pairs of sets:

1. A = {77, 88, 99}
   B = {88, 99, 110, 121, 132}
2. C = {a, e, i, o, u}
   D = {o, r, i, e, n, t}
3. E = {$\frac{1}{2}$, $\frac{1}{4}$, $\frac{1}{6}$, $\frac{1}{8}$}
   F = {$\frac{1}{3}$, $\frac{1}{5}$, $\frac{1}{7}$, $\frac{1}{9}$}
4. G = {0·04, 0·08, 0·12}
   H = {0·12, 0·16, 0·2}
5. I = {VII, VIII, IX}
   J = {X, XI, XII}

J Express as ordinary numbers:

1 4×10
2 4×10^2
3 7.5×10
4 7.5×10^2
5 7.5×10^3
6 0.456×10^3
7 $10^2 \times 10^2$
8 9.9×10^5
9 0.0007×10^4
10 109.08×10^3

K Express in standard form:

1 530
2 5,300
3 7,900
4 7,090
5 80,000
6 210,000
7 201,000
8 7 million
9 $1\frac{1}{2}$ million
10 99,000,000

L If it is 14.35 G.M.T. what is the time at places with the following longitudes?

1 30° E
2 30° W
3 60° E
4 75° W
5 0°
6 90° E
7 90° W
8 135° E
9 135° W
10 150° E

M Write in a shorter form:

1 $2a + a$
2 $2 \times a$
3 $3 \times 3 \times 3$
4 $3 + 3 + 3$
5 $a \times 2b \times 3c$
6 $a + 2a + 3a$
7 $3b + 0 + b + 5b$
8 $3b \times 0 \times b \times 5b$
9 $2 \times a \times 6 \times b$
10 $a \times 5 \times d \times 6$

N Change the following decimal numbers to binary numbers:

1 7
2 14
3 25
4 33
5 51

O Change the following binary numbers to decimal numbers:

1 101
2 1101
3 1011
4 10101
5 1101011

P 1 Use a set-square to draw the following capitals:

TIFH

2 Use a ruler and compasses to construct a 30° angle by the following method:
(a) Construct a triangle with each side measuring 5 cm.
(b) Bisect any one of the angles of the triangle.
Explain why this works.

3 Use a ruler and compasses to construct a 45° angle.

4 Use compasses and a set-square to divide a line into 6 equal parts.

5 Draw sets of 3 parallel lines 2·5 cm apart:
(a) using a ruler and compasses,
(b) using a ruler and set-square.